INTRODUCTION
TO
SECURITY

INTRODUCTION TO SECURITY

Gion Green
and
Raymond C. Farber

Security World Publishing Co., Inc.
Los Angeles, Calif. 90034

First Edition 1975

First Printing, January, 1975
Second Printing, August, 1975
Third Printing, August, 1976
Fourth Printing, December, 1976

ISBN 0-913708-20-8

Library of Congress Card Number: 74-31685

Security World Publishing Co., Inc.
2639 South La Cienega Boulevard
Los Angeles, California 90034

Printed in the United States of America

About the text . . .

INTRODUCTION TO SECURITY is a collaborative effort on the part of two security specialists with outstanding credentials in this important and growing field, uniquely qualified to present a comprehensive study of the history, scope and functions of security. Their text details the fundamental principles of physical protection, internal security, systems of defense, fire prevention and safety, with an overview of the career opportunities in security for business and industry, exemplified in such specific areas as retail, hospital, cargo and computer security, and security services.

About the authors . . .

Raymond C. Farber has been an internationally recognized spokesman and leader in the advancement of professionalism in security since 1964, when he founded *Security World* magazine. He is also the publisher of *Security Distributing & Marketing (SDM)* magazine, and a major list of authoritative security books and pamphlets. In 1967 he founded the first International Security Conference, now held in major U.S. cities three times annually, for which he continues to be the sponsor and General Chairman.

Gion Green gained his first orientation to security as a Major in the regular United States Marine Corps, with many assignments in legal investigation and courts-martial. He resigned his commission to become a consultant to business and industry on minority groups, later establishing his own management consulting firm specializing in minorities and related concerns. He has also been an executive with two major television networks. A graduate of Princeton, Green is a published author and seminar lecturer on security.

CONTENTS

Preface

PART I: INTRODUCTION

Anglo-Saxon Period • Bond of Kinship • Lord/Man Relationship • Social Structure • Shire-Reeve • Early Courts • Oath and Ordeal • Norman Period • William of Normandy • Henry II, the Law Giver • Courts of Assizes • Trial by Jury • Post-Norman Period • Magna Carta • Central and Circuit Courts • Coroner • Justice of the Peace • Privy Council • Expansion of Trade • Growth of Crime • Private Police • Henry Fielding • Jonathan Wild • Industrial Expansion • Laissez-Faire • Sir Robert Peel • Security in America • Colonial America • The New Nation • States' Rights • Police Power • Public Law Enforcement • Federal Agencies • Growth of Private Security • Pinkerton • Brinks and Burns • Railway Police • World War II • Security Today • L.E.A.A. • Future of Private Security

PART II: COMMON ELEMENTS OF SECURITY

Perimeter Security • Barriers • Fencing • Walls • Gates • Buildings• Barrier Protection • Inside the Perimeter • Parking Lots • Surveillance • Lighting • Kinds of Lighting • Interior Security • Windows • Burglary-Resistant Glass • Doors • Locks, Hinges, Keys • Emergency Exits • Area Doors • Old and New Buildings • Files • Safes • Vaults• Container Protection • Traffic Control • Visitors • Employee ID • Badges • Package Control • Vehicle Control • Inspections • Guard Patrols • Fire • By-Products of Fire • Classes of Fire • Extinguishers• Education in Fire Prevention • Fire Brigades • Evacuation • Planning Security • Security by Design • Principles

Why Employees Steal • What Employees Steal • Cash • Property • Methods of Theft • Controls • Internal Security • Management Involvement • Personnel Screening • Backgrounding • Polygraph • PSE • Morale • Survey Program • What Is Covered • Management Support • Communicating the Program • Supervision • Specific Controls • Audits • Cash Controls • Receipts and Deposits • Cash by

Exits • Scaffold Safety • Sanitation • Toilets • Washing Facilities•
Lunchrooms • Marking Hazards • Fire Protection Standards

PART III: CAREER OPPORTUNITIES IN SECURITY

The Emerging Security Function • Security Manager's Role • Inte-
gration of Security • Authority of Security • Relation to Other De-
partments • Functional Authority • Levels of Authority • Security
Administration • Planning • Organizing Security • Controls • Security
Department • Hiring • Supervision • Security Image • Departmental
Evaluation • Performance • Cost-Effectiveness • Personnel • Equip-
ment • Procedures • Applications in Business and Industry • Reduced
Losses and Net Profit • Career Opportunities • Profile of the Security
Manager • The Future in Security

Cost of Losses • Shoplifting • Methods • Who Shoplifts • Preventive
Measures • Surveillance • Mirrors • Signs • Displays • Check-Out •
Refunds • Detectives • Spotting the Shoplifter • Arrest • Prosecution
• Checks • Advantages and Disadvantages • Cash on Hand • Bad
Checks • Types of Checks • Returned Checks • Approvals • Examin-
ing the Check • Identifying the Customer • ID Equipment and
Systems • Cooperative Systems • Burglary • Merchandise • Cash •
Defenses • Physical • Alarms • Safes • Robbery • Profile of Robber•
Targets • Prevention • Cashroom • Opening Routine • Messengers •
Closing • Employee Training • Internal Theft • At the Cash Register •
Giveaways • Price Changing • Kickbacks • Refunds • Theft • Embezzle-
ment • Countermeasures • Screening and Supervision • Shopping Ser-
vices • Controls • Trash • Package Control • Employee Morale

Hospital Vulnerabilities • Failures of Management • Basic Security
Needs • A Hospital Security Program • Security Personnel • Patrols •
Entrances • Perimeter Openings • Security Posts • Command Post •
Interior Patrols • Visitor Control • Uniforms and Weapons • Badges
and Passes • Internal Controls • Employee Screening • Services, Cash
and Supplies • Insurance • Fire Prevention • Flammables • Electrical
Hazards • Vulnerable Areas • Inspection • Emergency Exits • Super-
vision • Fire Marshal • Alarm Systems • Planning and Training

Preface

The enormous growth of security in the past two decades has raised many questions that have only recently begun to come into focus. The heavy demands upon the resources of public police agencies created by the sharp increase in criminal activity has to some degree diverted their attention from traditional efforts in the protection of private property, leaving that responsibility more and more to the proprietor.

In turn, this has placed private security in the position of playing an increasingly active role in many areas of preventive law enforcement. To meet the current needs for security services, as well as the increased demands which can be foreseen in the future, private security must be prepared to evaluate its position and to improve the quality of its performance. Nothing less will meet the challenges faced by a free society in a changing world.

Every effort must be made to professionalize an industry, now in a state of change and dynamic growth, which plays such a significant role in our social as well as our economic structure. Universal standards of training and performance must be sought. Above all, security must search for and develop innovative personnel with managerial talents to assume the future leadership of this growing industry.

It is the purpose of this work to provide a broad overview of private security in its practical application, and to suggest certain theoretical approaches to some of its problems. Additionally, it examines security in a selected sample of various industries to provide a more comprehensive picture of the career opportunities the security field offers.

The book is designed, first of all, as an introductory text for those students who are interested in a career in security; for those presently active in the field who are pursuing additional academic knowledge; and for those students concentrating on business administration who wish to examine the basic principles of an organizational function which may soon become an organic part of every business. In the opinion of the authors, such a study is a requisite to an understanding of the needs of the modern business structure.

Security managers and others with a professional interest in the subject should also find the text of value. Its general review of fundamental principles should provide an effective yardstick against which to measure particular experience. And its detailed re-examination of security practices may prove useful to security professionals in evaluating existing programs and systems under their supervision.

PART I
INTRODUCTION

Chapter 1

HISTORICAL BACKGROUND OF SECURITY

Security, in its semantic and philosophical sense, implies a stable, relatively predictable environment in which an individual or group may pursue its ends without disruption or harm, and without fear of such disturbance or injury. The current preoccupation with security in an organizational sense, as a means by which this safety and stability can be achieved, is the end product of kaleidoscopic influences, many in conflict, which have brought us to our present state of perception. Throughout history, ideas about security and its role in society have been shaped by a wide variety of institutional patterns whose impact must be examined in their historical context for an understanding of the status of security today.

It is particularly in the history of our English antecedents that we can find the microcosm that in so many ways typifies the development of security systems throughout the western world. Nor can we overlook its overwhelming importance in our own cultural heritage.

SECURITY IN THE AGE OF FEUDALISM

Feudalism provided security for both the individual and the group to a very high degree. The Anglo-Saxons brought with them to England a predisposition to accept mutual responsibility for civil and military protection of individuals and the organizations made up by those individuals. They also brought with them a strong affinity for the feudal contract whereby an overlord guaranteed the safety of persons and property, and provided arms and treasures to vassals who administered the work of serfs bound to the land.

In a world of constant warfare between men of power, security could be found in no other way. Stability lay in the system and in the power and cleverness of the lord. Group security lay in group solidarity. Formal systems of security that developed over the years of the Middle Ages were largely confirmations of those systems toward which the people of this society had gravitated naturally.

The laws that followed were largely refinements of existing practices. As the population grew, older, more primitive customs became less effective. At the same time the changes in political structure, occasioned by the unification of England under a single ruler, diluted the responsibility of the communities for their own safety.

Post-Norman England, beginning with King John, saw the introduction of concepts declaring the supremacy of law over arbitrary edict, thus developing a base for confidence in the continuity of the system and its institutions. Above all there was a formal declaration of the individual rights and responsibilities as between the state and its subjects and between subjects themselves.

Anglo-Saxon Period (450 - 1066)

It is generally agreed that the Anglo-Saxons arrived in Britain in the middle of the 5th century. They consisted of a band of raiders sprung from three races of the Germans—the Angles (hence the word English), the Saxons, and the Jutes. They settled easily on the land, since tradition has it that a British (Briton) king ceded part of his territory to the newcomers on condition they defend him from the Picts and the Scots, thus creating a traditional security pact.

Anglo-Saxon communities grew up in various parts of England, although each of these settlements claimed sovereignty over its own immediate domain. It was not until the early 9th century that Egbert of Wessex, though never crowned king, was recognized as ruler of all England. His supremacy was short lived, however, and it remained for his descendants in the 10th century to reconquer those areas occupied by Viking invaders and become sole rulers of all England.

Thus, in some six hundred years, the raiders from Germany (*Anglo-Saxons*) had destroyed the Picts, driven the Scots further north, pushed most of the Britons west into Wales, and consolidated England under a single ruler. These Anglo-Saxon invaders were heathens, but they were by no means savages. They brought with them a stable social structure and a fairly advanced legal system.

The basic ties of society were kinship and the voluntary association between lord and man. From these two relationships came the basic social contract that, whatever the increasing complexities of society, was the bedrock of English intra-community relationships for centuries to come.

Bond of Kinship

Kinship was the bond on which the individual depended for security and mutual aid in all his affairs. It was the duty of kindred to avenge or exact compensation for an injury done to its members. At the same time, they were responsible for bringing the offender to justice to answer for his conduct and to contribute to the compensations and fines that adjudication might exact. In this relationship, kindred might extend to third and fourth cousins or beyond. It was the fact and acknowledgment of kinship that created the tie.

Lord/Man Relationship

Between lord and man, the pact was bilateral—a *quid pro quo.* In return for maintenance, weapons, horses, lands and treasures, man gave total unswerving loyalty to his lord. He would lose his life willingly rather than betray the vow of loyalty.

Both of these are examples of bonds of mutual security repeated in analogous ways throughout the western world over the centuries. Out of this predisposition to provide mutual aid and to accept responsibility grew the organizational structure that was to continue for several centuries.

Social Structure

Groups of ten householders were considered a unit called tithings; ten tithings constituted a "hundred," and several "hundreds" made up a "shire." This system, under 9th century law, was known as the "frank-pledge" and required that each man be responsible for the good behavior of the rest, that he produce any of the rest charged with an offense, or, alternatively, make restitution.

The shire was originally in the charge of an "ealdorman" (hence the modern alderman in the United States), a title later changed to earl in Britain. By the 10th century there were fewer earls, so each one had several shires under his administration. This necessitated the creation of the office of "reeve" of the shire, or "shire-reeve," the forerunner of the sheriff, to administer a single shire.

The sheriff was responsible for holding courts twice a year in every hundred in the shire. These courts constituted the local police organization in criminal matters, the sheriff's "tourn" or circuit being the shire courts held for a particular hundred.

Justice by Oath and Ordeal

In the exercise of law, the Anglo-Saxons used the oath or the ordeal. The defendant in most cases produced an oath of his innocence with the help of a fixed number of compurgators. But, if he could not find the prescribed number to attest to his innocence, or if the circumstances of his case were particularly suspicious, his case was tried by ordeal—the judgment of God. At this point the church took over the proceedings and the plaintiff decided whether the ordeal should be of water or of iron.

In these courts the assembled persons determined whether the accused should be allowed to produce an oath; it was not left up to the sheriff in charge. The courts might have been intolerable otherwise, since many of them were private courts that were established when the king had granted favorites the jurisdictional rights over their own lands or men. The procedure was the same in private courts as in others. The judgment as to the manner of the trial was still in the hands of the citizens, although any fines went to the lord of the estate.

The penalty for certain crimes such as treason, murder not in combat, and serious theft, was death; and by the 11th century a set of offenses of a grave nature were designated as the "king's pleas" and were rarely permitted in private courts. The judgment for most offenses, however, was satisfied by the payment of compensation to the persons harmed and a payment of a fine to the state or to the holder of the rights of private court. Failure or inability to pay would result in a period of servitude in the household of the plaintiff. Failure to appear before the court to answer a charge made one an outlaw.

Class Structure

Anglo-Saxon society was made up of three classes: nobles, churls, and slaves. The first two had a price or *wergild* set upon them. This was a prescribed amount to be paid to surviving relatives if some-one should kill one. Nobles had much the same wergild throughout England, but churls varied in the level of theirs. The slaves had no wergild but, in the event of their death by violence, a payment was due the master. Generally the churl's wergild was about one sixth that of a noble. Other compensations increased in much the same way according to rank. In the event of injury to a noble, the com-pensation to him and the fine payable to the court would both be higher than the levy against a similar injury to a churl. On the other hand, an offending noble paid larger fines for his offenses; higher rank had greater responsibilities. Noblesse oblige. It is inter-esting to note that among the injuries inflicted requiring compen-sation was the infringement of a man's right to protect his house-hold—a right referred to as his mund.

Origins of Law

The pattern of Anglo-Saxon life, or more particularly the man-ner in which it was regulated, shows important parallels to much earlier cultures. The earliest written law known is contained in The Code of Eshnunna (1950 B.C., *approx.*), the Laws of Lipit-Ishtar (1875 B.C., *approx.*), and the Code of Hammurabi (1750 B.C., *approx.*).

None of these are in any real sense codifications of the law of land; but they appear to be amendments or specific applications of existing common law, which is essentially composed of an understanding of long established customs and attitudes regarding intra-community relationships. They offer nothing in the way of abstract definitions of crimes or the laws concerning them. They are rigidly penal in form, devoting their full attention to listing the nature of the punishments for a miscellany of listed offenses.

With an eye to a certain kind of primitive equity the punishments may, in some cases, mirror the offense by treating the offender as he treated his victim or, in others, retaliate on the basis of injury-for-injury, such as cutting off the offending hand that injured the aggrieved party.

Implicit in such law was the notion that satisfaction between litigants could only be achieved by supervised, adjudicated, retal-iatory action. The state was in no way offended by such conduct between its citizens; but, to protect the weak from the strong, the state took a role in dispensing justice. There is no evidence that malefactors were brought to justice on the initiative of the state, except in cases of treason, subversion, or other such direct assaults upon the stability of the community. The courts handled as civil matters cases we now think of as criminal; and the court supervised the discharge of penalties, ranging from the payment of compensation to maiming or death by drowning.

In much the same way, the Anglo-Saxon structure of law grew out of those attitudes deeply ingrained in their tribal structure. The strong bonds of kinship were codified into law in the creation of tithings, which restated and formalized the ties and responsibilities between men in the same community. At the same time, there was little sense of community involvement in matters concerning private wrongs. Up until the 13th century criminal actions were instituted by appeals to the county court and they proceeded on private initiative rather than by any action of the crown. The idea that wrongdoing might injure not merely one person, but damage or endanger the entire community, was not developed until much later.

In most of this period crimes were looked on as acts of war—and feuds and private warfare were normal occurrences. Since hostility among the various English kingdoms was a constant fact of life until the end of the 9th century, private wars were simply consistent with the climate of the times. Authorities were concerned with reconciling antagonists in such disputes on established terms rather than with the exertion of legal processes. Ultimately, the notion that the "King's Peace" was antithetical to unlimited private warfare began to gain some attention around the end of the 9th century under the rule of Alfred, the Great.

Norman Period (1066 - 1199)

William of Normandy

In October of 1066 the Anglo-Saxon world found itself in new hands. William of Normandy had submitted his claim to the throne of England to trial by combat and with his victory at Hastings had himself crowned king. Invasion was nothing new, since for the last two hundred years the Viking wars had occupied the attention of the English. But William showed himself to be an able, if ruthless, administrator. His claim to the throne was soon generally accepted—and, in fact, by 1069 the English army was fig' his behalf.

The most significant aspect of his invasion and accession to the crown was that William never thought of himself as a conqueror but only as a legitimate heir. He was crowned with the ceremonial of old English kings and swore their oath at his coronation. The shire courts and hundred courts remained, and manorial or private courts continued as before.

Life in England was not radically altered—largely because William had no better system to impose. His native Normandy was disordered and fragmented and the social stability he found in England hardly indicated the need for great change. The old English aristocracy was replaced by Normans, and the estates of his enemies were seized and conferred on those loyal to him; but he made no attempt to redivide them. The continuity of land holdings was preserved.

William did, however, make certain important administrative changes whose trend was toward centralizing the governmental and judicial procedures. Law enforcement was administered through fifty-five districts which corresponded roughly to those shires previously in existence: The shire-reeves were no longer permitted to preside over courts, being replaced by the king's judges, who now assumed the "tourn" or circuit which the shire-reeves had been assigned.

The shire-reeves assumed administrative duties increasingly directed toward the preservation of order within their jurisdiction. The army was strategically garrisoned for purposes of defense, but it undertook certain rudimentary police duties such as the protection of roads and, at the request of the sheriff, the pursuit of outlaws. The idea that criminal offenses were an affront to the state was developing—along with the state's increasingly active role in bringing law breakers to account.

Henry II, The Law Giver

In the 12th century, under the rule of Henry II, the Law Giver, an important step was taken in establishing a line between those crimes deemed to be serious or felonious, such as murder, arson, robbery, false coinage and crimes of violence, and those lesser offenses which are judged to be misdemeanors. This clear statement by Henry advanced the earlier more informal treatment of the "king's pleas" to a formal definition requiring prescribed action by the courts.

Henry II can be truly said to have initiated the rule of law as distinct from the more arbitrary judgments of the Norman kings. In successive "assizes" or codes issued with the sanction of councils of barons and churchmen, he advanced a system of significant reforms.

The Assize of Clarendon in 1166 revived the old English system of mutual security or frank-pledge. No stranger was permitted to stay over in any place but a borough (or town), and he was permitted to stay there only for one night unless townsfolk would provide some guarantee of his good behavior. The list of such strangers was submitted to the justices.

This assize provided also the origin of trial by jury. Twelve lawful men of each hundred, with four from each township, were sworn to present those who were either known or reputed to be criminals within their district for trial. The jurors were thus not merely witnesses, but judges as well in determining the validity of the charge. This two-fold characteristic has remained in today's grand jury system, which still has the duty of presenting criminals for trial after hearing witnesses against them. In later years, under Edward I, witnesses acquainted with the facts at hand were added to the jury in each case—along with others who had no knowledge of the matter.

This assize additionally abolished the ancient practice of compurgation, in which an accused could be acquitted of all charges by the voluntary oath of his neighbors and kinsmen. For the next fifty years accused persons brought to trial by the grand jury were subjected to an ordeal, or the judgment of God. The Lateran Council of 1215 abolished the practice of the ordeal, leading the way to trial by "petty" juries.

Post-Norman Period

Magna Carta

The uprising of the Barons against King John ultimately resulted in the signing, also in 1215, of the Magna Carta—a document that proved to be of enormous significance in the history of England. It established the principle of due process by stating that no freeman could be arrested, imprisoned, deprived of his property, outlawed, exiled, or "in any way destroyed" except by "legal judgment of his peers or by the law of the land." It additionally established the important principle that the king is subject to the law.

The importance of this document cannot be overemphasized. Its significance was such that whenever the barons or, in later years, the Parliament felt the king was being arbitrary, they insisted on reaffirmation of the Magna Carta. This was done 44 times before the death of Henry V in 1422, and it has been used by lawyers frequently since.

These developments in the administration of the law, though precursors of the state's increased involvement in the welfare of its citizens, had no appreciable effect on the citizen's personal security. Walled towns, fortified structures, and citizens' committees were still the order of the day, and each person was primarily responsible for his own protection and the protection of his property.

The work of Henry II in reforming the judicial system by the introduction of juries, as well as his efforts to codify the law, was of great importance to English legal process—which was undergoing significant gradual changes that would not be completed until the 14th century. Three central law courts and a circuit court resulted.

Development of a Central Court System

The three central courts were concerned with common law; that is, law that had national application and was neither of merely local nor special application, such as the Ecclesiastical Courts. Originally, these courts had separate jurisdictions: The Court of the Exchequer dealt with tax and revenue cases; the Court of Common Pleas dealt with civil matters between citizens, and the Court of King's Bench dealt with a wide variety of matters but, most importantly, it considered criminal cases and exercised a general supervision over the lesser courts.

These courts had grown out of the King's Court, which, in turn, had been a virtually indistinguishable element of the King's Council, that body of aides and advisors by which he conducted the affairs of government.

The Council performed all the functions of government. It legislated the laws, administered them, and adjudicated cases growing out of them. By and large, the judgments of the Council were more equitable than those of the long established local and feudal (private) courts, and its influence grew steadily.

As the judicial work of the King's Council grew, it became evident that this specialized function could no longer be handled in a body of broad administrative and legislative duties, and the courts broke off from the Council as a separate body.

The actual operation of these courts was involved with the circuit system that pre-dated Norman England, when the sheriff was required to tour the county and preside at court sessions. William I had centralized this system by dispatching commissioners to oversee local matters.

Under the new system, local juries compiled information and presented cases involving serious crime to a circuit judge representing the King's Court. Such judges made decisions as to the handling of the case, as in a modern hearing; but they did not, at this time, make an adjudication.

By the 13th century, the increase in the jurisdiction of the King's Court had served to reduce the role of the old county courts and "courts of hundred" as well as feudal or private courts. But this created a difficulty for litigants. Jury trials were becoming common, and juries consisted of neighbor witnesses. If it was so determined by the circuit judge, all parties to the case, including the jurors, had to travel to Westminster, then the seat of government, for a hearing before the king's court! Obviously such a system needed revision.

A compromise was arrived at in the second Statute of Westminster in 1285. Cases in which an issue of law was involved would be determined by several judges sitting *in banc* in Westminster. Cases involving an issue of fact would be heard by juries. Writs and other paper work were presented to the court in Westminster for such initial consideration as was required to determine their manner of hearing.

In trials involving juries, a circuit judge visiting a county on his tour would hear the case, and the verdict of the jury would be taken. The case would then be returned to Westminster for judgment.

The courts in Westminster were therefore concerned with legal argument and the adjudication of cases in which the verdict had been rendered, while criminal and civil cases before juries were heard locally in the counties by what came to be known as the

Courts of Assize. Thus, under the rule of Edward I, the judicial reforms begun a century before by Henry II were effectively centralizing the administration of justice while still maintaining local involvement in the conduct of the litigation.

Coroners

Two new institutions arose to support the royal system. The administration of county affairs had been for several centuries in the hands of a sheriff. The power over local affairs concentrated in the hands of these men often led to oppression and corruption; hence *coroners* appear in the 13th century.

Four coroners were appointed by each of the new county courts to keep records for presentation to the visiting judges. Additionally, in overseeing the interests of the crown, coroners conducted inquiries into a variety of matters including the circumstances surrounding unexplained deaths. These inquests were held with the help of juries to determine presumptive facts for presentation to visiting judges if necessary; but the coroners did not at any time try cases.

Justices of the Peace

The other new development was the creation of the *justice of the peace.* A series of statutes in the 14th century, particularly those of 1361 and 1363 under Edward III, authorized these men, who were appointed by the crown, to hold sessions four times a year to hear and determine criminal cases—thus relieving the visiting judges of all but the most serious.

Chancery Courts

When the common law courts separated from the King's Council, most of that body's judicial powers went with it. The residual judicial powers were vague; but people who, through defects in the law or its administration, or through their poverty, felt they could not get justice in the courts, petitioned the king or his council for equitable relief. Such petitions were handled by the lord chancellor. Any resultant decrees were made as an act of the Council, of which the lord chancellor was an important member.

In 1474 the lord chancellor began to make decrees on his own cognizance, and the Court of Chancery came into being. This court was supplementary to the common law courts in that it could give remedies where common law provided none. It was essentially a court of equity and, since no jury was involved, adjudication was solely at the discretion of the lord chancellor, who determined disposition of the matters before him as good conscience dictated.

The Privy Council and the Court of Star Chamber

Even with the separation of the Court of Chancery from the King's Council, the king exercised certain judicial powers. Upon the accession of Henry VII to power by force of arms, the Council was used as an instrument to cope with powerful subjects difficult to deal with by ordinary judicial means.

The Council was further reorganized into advisors who accompanied the king in the conduct of affairs of state, and those who handled administrative work in Westminster. Those counsellors with the king became known as the Privy Council, while those in Westminster—who found that much of their work was judicial—became known as the Court of Star Chamber (since they sat in a chamber decorated with a starred ceiling). This court had jurisdiction over matters of forgery, riots, perjury, libel, conspiracy and fraud—and it had the power to inflict any punishment short of death. Its process was not controlled in the same manner as that in the courts of common law. It was summary in nature, and its judgments were frequently arbitrary and cruel. It often forced confessions and sometimes resorted to physical torture.

But a larger danger was its use to enforce proclamations of the king. Magna Carta had established the principle that the king was subject to the rule of law; but if the king was allowed to actually govern by edict and proclamation, he could subvert the common law that Parliamentarians held to be the supreme authority. Chief Justice Edward Coke ruled in 1610 that the king (James I) might not create new offenses by proclamation. But as long as the Court of Star Chamber existed, such proclamations would be enforced whatever their validity. Hence, in 1641, in the reign of Charles I, the Court of Star Chamber was abolished by an aroused Parliament.

Development of Law Enforcement

Concurrent with judicial reform were those measures specifically aimed at the enforcement of public order. The Statute of Winchester (also known as the Statute of Westminster) in 1285 revived and reorganized the old institutions of national police and national defense. It described the "duty of watch and word," which is to say that every man had the duty to pursue and bring to justice felons whenever "hue and cry" was raised.

Every district was made responsible for crimes committed within its bounds; the gates of all towns were required to be closed at nightfall, and all strangers were required to give an account of themselves to the magistrates. Brushwood and other concealments were to be cleared for a space of two hundred feet on either side of public highways to protect travelers against attack by robbers.

Under Edward III, regulations were brought about allowing boroughs extended rights to enact their own ordinances. Attempts were made to control vice and crime, and ordinances to that end were enacted locally. Since organized agencies for the enforcement of such laws were virtually nonexistent, these efforts had a limited success. Night watches and patrols were privately established, however, for the protection of citizens from direct assault.

Subdivision of Land

The development of systems of protection and enforcement appeared to come with greater rapidity and sophistication from the 14th through the 18th century; but the seeds for this development had been planted during the social revolution that heralded the end of the remaining elements of the feudal structure, in the latter half of the 13th century.

The basis for this change was the subdividing of the land. The national wealth was increasing, and the country gentry and the substantial yeomanry in increasing numbers became land owners. Tenants of the great barons now received under-tenants, who paid to the tenant what he would be paying to his lord. The barons, though still receiving full rents and service according to which they had originally granted their lands in fee, were nonetheless disturbed when they saw the profits to be made in which they had no participation. The value of their estates had substantially increased because of subdivision and higher cultivation of the land.

Thus, under pressure from the barons, Edward I, with the concurrence of Parliament, enacted the Third Statute of Westminster in 1290, which forbade the practice of subfeudation. It specified that anyone buying feudal rights from a tenant was henceforth not the tenant of the tenant who sold the rights, but of the baron of that estate.

Its result, however, was quite opposite from its intent; and the statute served to promote the transfer and subdivision of land rather than to hinder it. The original tenant was obliged to keep only so much of his contracted service as would enable him to discharge his obligation to his lord; the rights to the remainder could be sold to a new tenant. The estates thus created might be small, but the practice created a growing class of lesser gentry and freeholders, and hastened the demise of feudalism.

EXPANSION OF TRADE

With the increased interest and success in voyages of exploration seeking new markets and trade routes, means were opened by which trade of all kinds could be expanded enormously.

This period of exploration was roughly concurrent with acts of enclosure that prevented common grazing, or other use of land that feudal lords had consolidated by driving out small holders. Whether this was in reaction to sub-tenancy evasions or an effort to create a base for the accumulation of raw materials such as wool and other products is difficult to determine. The effect was the same.

Displaced tenants migrated to the cities in great numbers. Urbanization of the population created conditions of considerable hardship. Law enforcement and existing control systems were unable to cope with the changing conditions, and poverty and crime increased rapidly. Riots were reported as early as 1450.

Efforts Toward Communal Security

By the beginning of the 16th century the social patterns of the Middle Ages were being broken down. No public agencies existed that could restrain the mounting wave of crime and violence, and no agencies existed that could alleviate the root causes of the problem.

Different kinds of police agencies were privately formed to deal with the crime problem of the day. Individual merchants hired men to guard their property. Merchant associations also created the *merchant police* to guard shops and warehouses. Night watchmen were employed to make their rounds. Agents were employed to recover stolen property. *Parochial police* were employed by the people of the various parishes into which the city was divided.

Attention then turned to the reaffirmation of laws to protect the common good. Although the Court of Star Chamber had been abolished in 1641, its practices were not officially proscribed until 1689, when Parliament agreed to crown William and Mary, if they would reaffirm the ancient rights and privileges of the people. They agreed, and Parliament ratified the Bill of Rights—which for all time limited the power of the king as well as affirming and protecting the inalienable rights of the individual.

By 1737 a new aspect of individual rights came to be acknowledged; for the first time tax revenues were used for the payment of a night watch. This was a most important development in security practice, since it was a precedential step that established for the first time the use of tax revenues for security purposes.

Eight years later, Parliament authorized a special committee to study security problems. The study resulted in a program employing various existing private security forces to extend the scope of their protection. The resulting heterogeneous group, however, was too much at odds and never proved to be effective in providing any satisfactory level of security.

A Professional Police Force Proposed

In 1748, author and magistrate Henry Fielding proposed a permanent, professional, and adequately paid security force. His invaluable contribution was a foot patrol to make the streets safe, a mounted patrol for the highways, the famous "Bow Street Runners" or special investigators, and police courts.

It is interesting to note here that, among other plays, essays and novels (including the unforgettable *Tom Jones*), Fielding wrote an ironic novel called, *Jonathan Wild—The Story of a Great Man.* Its hero, Jonathan Wild, was a real person in 18th century London, perhaps the most notorious fence, thief-taker and master criminal of modern times. He was so real, in fact, that an account of his activities occupies eight pages of a staff report prepared in 1972 for the Select Committee on Small Business of the United States Senate, more than 200 years after he was hanged at Tyburn.

How is it that the spectre of Jonathan Wild still haunts those bodies charged with finding means to minimize crime? In many ways Wild's career typified the problems of security—or, more specifically, theft control—in the 18th century. He was also an archetype that suggested patterns still with us, unresolved, as the following portrait from the Select Committee's report indicates.

Jonathan Wild—Portrait of a Master Criminal

"Wild, a bucklemaker by trade, first came to official attention as a debtor, for which he was committed to the Wood Street Compter. Unlike many debtors of his time, he managed not only to survive but also to endear himself to enough officials to remove himself to the more comfortable part of the prison and become a trusty.

"In 1712, therefore, when an act for the relief of insolvent debtors was passed, Wild was one of the first to be granted release and pardon. At this point it is said that a woman introduced him to crime, or at least to the London 'street scene,' as a way of life. More likely is that the corruption of officials such as jailees and magistrates which he observed while in the compter had convinced him that crime was a most lucrative field in which to make one's way in the world.

"Wild began, in fact, by working in close association with the somewhat notorious under city-marshall Charles Hitchen who introduced Jonathan to the specific enterprises of 'recovering' property and extorting allegiance from thieves, as well as the fine art of bribery. Hitchen's tactics were, however, something less than subtle; and whether it was because Wild found these too crude or because he felt he had learned enough from the city-marshall, he struck out on his own in 1714 with the advertisement quoted below.

" 'Lost on Friday evening 19th March last, out of a Compting House in Derham Court in Great Trinity Lane, near Bread Street, a Wast Book and a Day Book; they are of no use to anyone but the Owner, being posted into a ledger to the day they were lost. Whoever will bring them to Mr. Jonathan Wild over-against Cripplegate-Church, shall have a guinea reward and *no Questions asked.*' (*Daily Courant*, 26 May 1714)

"In nine short years, Jonathan had built from a modest lost property business an organization which exerted almost total control over London's underworld. From recovering 'lost' property, Wild first branched out into the enterprise of thief-taking (i.e., locating, 'arresting,' and actively pursuing the conviction of thieves).

"This activity evinced the spirit of public service by which Wild chose to be known and served to further allay any suspicions that his lost property office might be founded on dealings of a distinctly unsavory nature. It had the additional advantage of being extremely lucrative. When he began, reward for bringing a thief to justice was 40 Pounds, being changed to 140 Pounds in 1721 (on the basis of a suggestion Wild himself made to the Privy Council in 1720). The combination of thief-taking and recovering property, then, made him not only especially knowledgeable but also financially secure.

"His interests began to spread. He bought a sloop and kept busy a smuggling trade between London and the Continent; he made arrangements with the agent who contracted the ships used to transport criminals to North America so that he was immediately notified of those who returned prematurely and could commit them to Newgate (at 40 Pounds per head) for the reward; he employed specialists to instruct bands of his pick-pockets and thieves in the fine art of 'mixing well' in high places so that they could be discharged to the same in pursuit of valuable booty.

"Jonathan began also to upgrade his personal life style, moving to Old Bailey and purchasing a private coach and team. At the time of his death, a conservative estimate of his fortune could be put at over $1-1/2 million and there was no question that little or nothing happened in and around the metropolis that Jonathan didn't know about, have some interest in, or first approve. Even more significant, there seemed to be very little indeed that could happen without his approval and manage to escape his wrath as well.

"Because Jonathan Wild was such an extraordinary criminal, it is easy to lose sight of the fact that, first, he was at base a receiver, and second, that his whole organization was geared to facilitate that primary enterprise. Thus, thief-taking and bribery were his means of exercising control over the underworld which did his bidding. Smuggling provided him with outlets for 'lost' property which he could not safely dispose of in England. Similarly his activities with regard to transportees gave him a source of operatives to carry out theft as well as a pool of patsies to whom he could ascribe crimes committed by his more favored 'associates.' His elaborate training of thieves allowed him to exert quality control over his employees as well as providing him with the opportunity of endearing himself to 'the quality' by 'recovering' their property.

"But if we, possessing an historical perspective, can be deceived into thinking Wild was something more than a very clever fence, his contemporaries were overwhelmed by that impression. Wild was toppled because he was Wild, 'a perfectly new scene' in DeFoe's words. And when all the indignation surrounding his death had died down, there was little doubt that it was the man and not the machinations which were objected to. Wild's life, as we shall see, 'taught no lessons, brought no reforms, and alleviated no suffering.' " [1]

Property Crimes and Law

For many centuries the English common law almost totally ignored the receiver of stolen goods. The focus of the law was upon the receiver of men, not goods. To give aid and assistance to a fleeing felon made a man guilty as an accessory after the fact to that felon's crime. But mere receiving of stolen goods with knowledge did not make the receiver an accessory after the fact in any case.

Perhaps this attitude of the common law can be explained, in part, by the relative unimportance of dealings in stolen property in the early stages of the development of the law of crimes. Until the 17th century the amount of movable property available for theft was probably limited and opportunities to dispose of this property, other than by personal consumption, rather restricted. Lacking a professional police force, the attention of the community, and the law, was therefore primarily directed towards apprehending offenders rather than tracing and recovering stolen property. The victims of property crimes were left to rely upon their own ingenuity, bolstered by several shaky legal remedies, to secure the return of their plundered goods and chattels.

It was only in the late 17th century that the legislature moved for the first time to combat the problem of the receiver of stolen goods. In 1691, under a statute by William and Mary, the receiver was made subject to prosecution as an accessory after the fact. The preamble to this statute recited that many "thieves and robbers are much encouraged to commit such offences, because a great number of persons make it their trade and business to deal in the buying of stolen goods."

However, the legislative desire to proscribe this trade and business was largely thwarted by the rule that no prosecution could be brought against the receiver until the principal felon, the thief, had been brought to justice. Despite a subsequent statutory modification of this rule in the early 1700's which permitted the receiver to be punished as for a misdemeanor if the principal felon could not be taken, the tradition remained throughout the 18th and early 19th century that the receiver was an accessory rather than a principal to the crime.

INDUSTRIAL EXPANSION

The Industrial Revolution began to gather momentum in the latter half of the 18th century. By 1801, the poet William Blake of apocalyptic vision was writing disapprovingly of "these dark, Satanic mills." Like the migrations off the land two hundred years earlier, people again flocked to the cities—not this time pushed as they had been by enclosure and dispossession, but lured by promises of work and wages.

The already crowded cities were choked with this new influx of wealth seekers. What they found was the hours were long, wages were miserly, and men and women—even tiny children—worked in unsafe factories. Disease periodically swept the crowded quarters. Family life, heretofore the root of all stability, was virtually destroyed in this environment. Thievery, crimes of violence, juvenile delinquency were the order of the day. All the ills of such a structure, as we can see in analogous situations today, overtook the industrial centers.

Adam Smith and Laissez Faire

Adam Smith had, in 1776, gained a large and appreciative audience for his *Wealth of Nations.* In it he contended that labor was the source of wealth; and it was by freedom of labor, by allowing the worker to pursue his own interest in his own way, that the public wealth would best be promoted. Any attempt to force labor into artificial channels, to shape by laws the course of commerce, to promote special branches of industry in particular countries, would be not only wrong to the worker or merchant, but harmful to the wealth of the state.

With general acceptance of such statements of laissez faire, it is small wonder that little was done to alleviate the growing problems. Crime rates spiralled. Counterfeiting was so common that at one time it was estimated that more counterfeit money than government issue was in circulation. Over fifty false mints were found in London alone.

Attempts to Control Crime

The response to such a high crime rate was predictable. Penalties were increased to deter potential criminals. At one time over 150 capital offenses existed. These ranged from picking pockets to serious crimes of violence. No visible decline in crime resulted. It was for all purposes a "society which lacked any effective means of enforcing the criminal laws in general. A Draconian code of penalties which prescribed the death penalty for a host of crimes failed to balance the absence of efficient enforcement machinery."[2]

Private citizens resorted to carrying arms for protection, and they continued to band together to hire special police to protect their homes and businesses.

Sir Robert Peel and the Metropolitan Police

In 1822, Sir Robert Peel became Home Secretary. He had an abiding interest in creating a strong, unified professional police force. This interest had emerged earlier when, as Secretary for Irish Affairs, he had reformed the Irish constabulary—members of which were thereafter referred to as "Peelers." As Home Secretary, Peel initiated the criminal law reform bill and he reorganized the metropolitan police force, also referred to as "Peelers" or, more commonly, "Bobbies." He also made efforts to decentralize police efforts and to develop the responsibility of each community for its own security.

Based on Peel's thoroughgoing reforms and revisions, the metropolitan police force became a model for law enforcement agencies in years to come. Modern policing was born with the "Bobby."

Unfortunately, not all of Sir Robert's efforts met with success. Neither the Police Act of 1835, establishing city and borough police forces, nor the County Act of 1839, setting up county police, nor various other acts passed in mid-century, created adequate police operations.

Private guard forces continued in use to recover stolen property, as well as to provide protection for private persons and businesses.

SECURITY IN AMERICA

Colonial America

Security practices in the early days of colonization followed the patterns that colonists had been familiar with in England. The need for mutual protection in a new and alien land drew them together in groups roughly analogous to the "hundreds" of bygone years.

Mutual protection and mutual accountability characterized the various groups. A semi-military flavor additionally created guard posts and occasional patrols. Sheriffs were elected as chief security officers in colonial Virginia and Georgia; and constables were appointed in New England.

As the settlers moved west in Massachusetts, along the Mohawk Valley in New York, and into central Pennsylvania and Virginia, the need for protection against hostile Indians was the principal security interest. Settlements generally consisted of a central fort or stockade surrounded by the farms of the inhabitants. If hostilities threatened, an alarm was sounded; and the members of the community left their homes for the protection afforded by the fort, where all able-bodied persons were involved in its defense.

Security in the New Nation

The American Revolution sprang from abuses similar to those which had created demands for reform in England, but which were aggravated in the colonies by frustration over their inability to get adequate relief. The Declaration of Independence delineates this sense of frustration and its attendant outrage with the greatest eloquence. But, perhaps more importantly, it declares a philosophy of government based on the rights of man, to secure which, "Governments are instituted among men, deriving their just powers from the consent of the governed."

Such an ideal can only promote the highest security. If indeed a government does govern with the concurrence of the governed, the continuity of its institutions is assured, and its stability creates an environment wherein security is a significant by-product.

The Fourth Amendment of the Constitution expresses this need by guaranteeing that "the right of the people to be secure in their persons, houses, papers and effects against unreasonable searches and seizures, shall not be violated."

In the Fifth and the Fourteenth Amendments, important assurances of security are given when it is declared that neither the government of the United States nor the governments of the states may deprive a citizen of "life, liberty or property without due process of law."

Post-Revolution Police Efforts

After the revolution, the United States made sporadic attempts to establish adequate security systems, but no unifying philosophy or principle was developed that could act as a guide in their formation. The absence of underlying principles for police and security was further aggravated by the multiplicity of overlapping jurisdictions in the new nation—yet these were an inevitable and generally desirable result of the principles upon which the country was founded.

States' Rights

The Constitution was carefully drawn and ratified as an instrument threading a narrow line between the rights of the several states and the power vested in the federal government. The fear of excessive and perhaps arbitrary power in a central government, like that experienced under the English crown, persuaded the Constitutional Congress to provide the federal body only those powers required to administer the broad needs of the land—and to delegate the rest to the individual states.

The issue was, of course, the cause of battle lines being drawn in the Civil War; but it has always, before and since, been a subject of much debate and discussion. Just where, and in what way, the police powers of the federal government should supersede those of other jurisdictions within its whole has been defined slowly over the years by numerous cases argued before the Supreme Court. The line of separation defining these powers may always be in flux; certainly it is subject to re-examination to this day.

Police Power

Police power is bestowed upon the state to permit interference with normal personal and property rights in the interest of conducting government in the general public interest. Implicit in such necessity is the intervention of government in the settlement of disputes, collection of revenues, enforcement of criminal law, the regulation of industry, commerce and agriculture, the administration of agencies providing for the public safety, etc.

In this regard, Chief Justice Roger B. Taney said in 1847, "What are the police powers of a state? They are nothing more or less than the powers of government inherent in every sovereignty to the extent of its dominions. That is to say—the power to govern men and things within the limits of its domain."

The exercise of various police powers through criminal laws or administrative action may, like the taxing power, interfere with property; but it does so not to collect revenue, but to promote the public welfare. Justification for such action is derived from common law in such maxims as *"salus populi suprema lex"* (the public safety is the supreme law).

The balance between the police powers of either state or federal government and the guarantees of due process embodied in both the 5th and 14th Amendments have also been the subject of much attention by the courts. The end result of such issues was that, although the exercise of a broad range of police powers was affirmed in various jurisdictions, the fragmentation created by these entities made for difficulty in the formation of efficient police (used in its narrowest sense) systems.

Today, there are roughly 40,000 police jurisdictions in the United States. Police efficiency and effectiveness must vary considerably throughout the country.

Establishment of Public Law Enforcement

The development of police and security forces seems to follow no predictable pattern other than that such development was traditionally in response to public pressure for action.

Outside of the establishment of night watch patrols in the 17th century, little effort to establish formal security agencies was made until the beginnings of a police department were established in New York City in 1783. Detroit followed in 1801, and Cincinnati in 1803. Chicago established a police department in 1837, San Francisco in 1846, Los Angeles in 1850, Philadelphia in 1855, and Dallas in 1856.

New York, influenced by the recent success of the police reforms of Sir Robert Peel, adopted his general principles in 1833. By and large, however, police methods in departments across the country were rudimentary; departments as a whole were inefficient, ill-trained, and corrupt.

The Spoils System

The declaration of the spoils system by Andrew Jackson had, if not created, at least articulated the system by which the party in power showered preference of all kinds, including police jobs, on its loyal followers. Such disregard for professionalism in favor of political expediency served only to decrease the already low public esteem for such agencies.

Not until the passage of the Civil Service Act in 1883 was any effective attempt made to upgrade the level of police protection, in spite of rising urban crime rates and frequent revelations of police corruption and inefficiency.

Federal Law Enforcement

In 1864, the U.S. Treasury Department organized its law enforcement agency—the second such federal service to be formed. The Post Office Department had set up its investigative service in 1828.

In 1870, the United States Department of Justice began operations, and the Border Patrol was formed in 1882. Although the Department of Justice organized an internal agency to investigate federal crimes as early as 1908, it did not effectively blossom into the national police agency known as the F.B.I. until 1932.

The Growth of Private Security

Allan Pinkerton

In 1850, Allan Pinkerton, a cooper from Scotland and the Chicago police department's first detective, established one of the oldest, and currently the largest, private security forces in the United States. The agency undertook to provide security and conduct investigations of crimes for various railroads and industrial concerns.

The services this agency provided were important to its clients largely because public law enforcement agencies were either inadequate to the job, or because their jurisdictions were such that they could not take necessary action. At the outset of the Civil War, General George McClellan, formerly of the Illinois Central Railroad, and a client of Pinkerton's, hired the agency as an intelligence gathering arm of the Union Army, since the army had no espionage apparatus at that time.

Brinks and Burns

In 1889, Brinks, Incorporated, started in business as a private service to protect property and payrolls. The William J. Burns, Inc., Detective Agency was formed in 1909 and became the sole investigating agency for the American Banking Association. For all intents and purposes, Pinkerton and Burns were the only national investigative bodies concerned with non-specialized crimes in the country until the advent of the F.B.I.

From the 1870's only private agencies had provided contract security services to industrial facilities across the country. In many cases, particularly around the end of the 19th century and during the depression in the 1930's, the services were, to say the least, controversial. Both the Battle of Homestead in 1892, in which workers striking that plant were shot and beaten by security forces, and the strikes in the automobile industry in the middle 1930's, are examples of excesses from overzealous security operatives in relatively recent history.

Railway Police

During the late 19th century, various Railway Police Acts enacted by most states authorized the railroads to establish their own security forces with full powers to police equipment and property. This was essential for the security of the railroads, since their property was extended through so many jurisdictions that security would have been impossible without a proprietary force in the absence of any federal agency empowered to cut across the various jurisdictions involved.

The growth of private railroad police in the early 20th century was huge. By 1914, over 12,000 of them were in operation, and by 1921 the Association of American Railways provided a coordinated security service to all of its members.

World War II and After

National security caused an even more dramatic development in the use of private protective forces during World War II. Before the end of the war, more than 200,000 men and women were sworn in as security personnel. Their training was undertaken by local police, but whatever level of competence they achieved, they were thrust into every conceivable kind of security position. Ultimately they were assigned to every facility across the country which was even remotely connected with production for the war effort.

Thus, in the post-war period, not only had private security personnel established their effectiveness in vast new areas, but they constituted a large work force, looking for jobs, trained and ready to usher in the new era of modern security.

Security Today

Like the threat to security inherent in World War II, the power politics of the Cold War years continued to bring the need for improved, more sophisticated security agencies and operations into sharp focus. The initial motivation in this increasing interest had been in protecting the nation against acts of sabotage and the theft of highly classified information, but the steady increase in crimes of violence as well as crimes against property served to direct attention to the need for countermeasures to combat this internal problem.

As a result of this attention to the growing problem of crime, both public and private security forces began to spread. In the last twenty-five years, public law enforcement personnel have increased by over 33 percent. And, while the statistics vary substantially from source to source, it would appear that employment in private security has increased by about 20 percent in the same period. By 1974 there were over 400,000 men and women employed in the private security industry, with just about as many in public law enforcement.

Federal Government's Role

Of the many forces acting to affect the business of security, the federal government seems at the moment to be the most significant. Although a breakdown of expenditures for crime prevention by all levels of government shows that about 75% was expended by local governments, with state and federal shares equal to about 12% each, the federal government's role is increasing significantly.

The L.E.A.A.

Basic in this deepening involvement in crime prevention at the federal level is the Law Enforcement Assistance Administration (LEAA), which is funded to encourage cooperation between local law enforcement agencies and to promote research, development programs and studies generally to improve the criminal justice system.

Established by the Omnibus Act of 1968, the LEAA grew out of the President's Commission on Law Enforcement and Administration of Justice, established in 1965. Its principal role is to allocate funds to state and local governments for the improvement of law enforcement procedures, but it additionally budgets funds for the collection and dissemination of crime statistics and, through its National Institute of Law Enforcement and Criminal Justice, develops new approaches, techniques, systems, equipment and devices designed to improve law enforcement.

This program, with nearly $1 billion budgeted annually, should begin to show some positive results in upgrading the effectiveness of local law enforcement agencies. However, it can be expected that these local agencies will largely concentrate on dealing with serious and violent crimes against persons, leaving much of the responsibility for the protection of property to private interests.

The Future of Private Security

If the trends suggested here continue, the growth of private security should increase sharply. As it is, expenditures on private security in 1974 amounted to almost $5 billion, a figure projected to grow at from 8 to 10% annually to a total of $6.35 billion in 1978. Of this growth, contract guard and security services reflect an even more rapid rate of expansion than in-house security organizations.

The reason for this development in all sectors of the security field is not difficult to identify. The crime rate is up in every area reported in the F.B.I. Index of Crime. The greatest escalation is in crimes related to property, with the most significant increases in the Robbery and Larceny $50.00-and-Over categories.

The Department of Commerce estimates that the annual national cost of crime to American business is nearly $16 billion; the department considers this estimate to be conservative.

Here, then, is the state of society today. Faced with alarming increases in crime on the one hand, and fragmented public law enforcement agencies on the other, it has turned increasingly to private security for protection—a trend that seems destined to continue.

The government has in no way discouraged this thinking. On the contrary, in its report on *The Economic Impact of Crimes Against Business,* the Department of Commerce concluded:

"Although businessmen are spending $3.3 billion[3] in private prevention efforts, there are indications that these efforts may be too passive. For example, insurance representatives claim that businessmen often take minimal precautions against robbery and burglary once they have obtained insurance. Also, any of the sources used for this study indicate that businessmen generally suffer from a lack of interest in either the extent of the problem or protective measures presently available.

"Thus, in a sense, businessmen may be tending to view criminal acts and rising crime rates as 'society's' problem. Unlike individuals who take stringent measures to protect themselves from criminal acts out of fear of bodily harm, businessmen tend to think of crime as something over which they have no control and from which they should be protected by others.

"However, even the scant figures in this report indicate that crime against the business sector is a matter of pressing concern. Businessmen should start to view crime just like any other component of business cost. Crime losses are costs. Cutting crime losses should receive the same sort of constant aggressive attention devoted to cutting labor costs, facility costs or any other costs that sap profits. This involves careful review of such things as the sufficiency and quality of present protective efforts and techniques, prosecution practices, and employee selection and training."

SUMMARY

Security, as we have seen, finds its sturdiest historical roots in both the concepts of justice and the social structure of early England. It is possible to trace the evolution of ideas of law and of mutual security through the Anglo-Saxon period to the dramatic developments of the Middle Ages, from the real origins of the rule of law under Henry II to the "giant step" embodied in the Magna Carta, and from the changes in the role of courts to the evolving awareness of the need for a more effective police function.

The awareness *was* a long time coming. Serious attempts to control crime in ways that went much beyond the mutual security of the Anglo-Saxons are not really seen until the 19th century, when the first professional police forces came into being.

The concepts of modern security and of modern police systems, then, are really quite new. It is important to remember, as we study the sophisticated principles and practices of security today, that public enforcement of criminal law in anything like its present form has been in existence only slightly more than a hundred years; and private security forces, whose history is longer, have only recently come into wide general use.

PART II
COMMON ELEMENTS
OF SECURITY

Chapter 2

PHYSICAL
SECURITY

The cause of security can be furthered simply by making it more difficult (or to be more accurate—*less easy*) for criminals to get into the premises being protected. And these premises should then be further protected from criminal attack by denying ready access to interior spaces in the event that exterior barriers are surmounted by a determined intruder.

This must be the first concern in security planning.

True, every security program must be an integrated whole—and each element must grow out of the specific needs dictated by the circumstances affecting the facility to be protected. But the first and basic defense is still the physical protection of the facility. Planning this defense is neither difficult nor complicated, but it requires meticulous attention to detail.

Whereas the development of anti-embezzlement systems, or even the establishing of shipping and receiving safeguards, requires particularized sophistication and expertise, the implementation of an effective program of physical security is the product of common sense and a lot of leg work expended in the inspection of the area.

Physical security concerns itself with those means by which a given facility protects itself against theft, vandalism, sabotage, unauthorized entry, fires, accidents and natural disaster. And in this context a facility is a plant, building, office, institution, or any commercial or industrial structure or complex with all the attendant structures and functions that are part of an integrated operation. An international manufacturing operation, for example, might have many *facilities* within its total organization.

The first step toward increased security, then, must be to minimize or control access to the facility. Such control is necessarily dependent upon the nature and the function of the facility and cannot ever interfere with its operation. It is theoretically possible to *completely* seal off access to a given operation; but it would be difficult to imagine how useful the operation would be in such an atmosphere.

Certainly no commercial establishment can be open for business while it is closed to the public. A steady stream of outsiders, from customers to service personnel, is essential to its economic health. In such cases the security problem is to control this traffic without interfering with the function of the business being protected. Isolated manufacturing facilities must also provide for the traffic created by the delivery of raw materials, the shipping of fabricated goods, services and, of course, the labor force—which may be operating in several shifts.

All such traffic tends to compromise the physical security of the facility; but security must be provided, and it must be provided appropriately for the operation of the facility to be served.

PERIMETER SECURITY

The perimeter will usually be determined by the function and location of the facility. An urban office building or retail enterprise will frequently occupy all the real estate where it is located; in such a case, the perimeter may well be the walls of the building itself. Most industrial operations, however, require yard space and warehousing even in urban areas. In that case the perimeter is the boundary of the property owned by the company. But in *either* case the defense begins at the perimeter—the first line which must be crossed by an intruder.

Barriers

Natural and structural barriers are the elements by which boundaries are defined and penetration is deterred.

Natural barriers comprise those topographical features that assist in impeding or denying access to an area. They may consist of rivers, cliffs, canyons, dense growth, or any other terrain or feature that is difficult to overcome. Structural barriers are permanent or temporary devices such as fences, walls, grilles, doors, roadblocks, screens, or any other construction that will serve as a deterrent to unauthorized entry.

It is important to remember that structural barriers rarely, if ever, prevent the penetration. Fences can be climbed, walls can be scaled, and locked doors and grilled windows can eventually be bypassed by a resolute assault.

The same is generally true of natural barriers. They almost never constitute a positive prevention of intrusion. Ultimately, all such barriers must be supported by additional security, and most natural barriers should be further strengthened by structural barriers of some kind. It is a mistake to suppose that a high, steep cliff, for example, is by itself protection against unauthorized entry.

Fencing

The most common type of fencing normally used for the protection of a facility is chain link, although barbed wire is useful in certain permanent applications, and concertina barbed wire is occasionally used in temporary or emergency situations.

Chain link fencing should meet certain specifications developed by the Defense Department in order to be fully effective.

It should be constructed of a wire of a number 11 or heavier gauge with twisted and barbed selvage top and bottom. The fence itself should be at least seven feet tall and should begin no more than two inches from the ground. If the soil is sandy or subject to erosion the bottom edge of the fence should be installed below ground level. The fence should be stretched and fastened to rigid metal posts set in concrete, with such additional bracing as may be necessary at corners and gate openings. Mesh openings should be no more than two inches square. The fence should, additionally, be augmented by a top guard or overhang of three strands of stretched barbed wire, angled at 45° away from the protected property. This overhang should extend out and up far enough to increase the height of the fence by at least one foot, to an overall height of eight feet or more.

To protect the fence from washouts or channeling under it, culverts or troughs should be provided at natural drainage points. If any of these drainage openings are larger than ninety-six square inches, they too should be provided with physical barriers that will protect the perimeter—without, however, impeding the drainage.

If buildings, trees, hillocks, or other vertical features are within ten feet of the fence, it should be heightened or protected with a Y-shaped top guard.

Barbed Wire

When a fence consists of barbed wire, a 12 gauge, twisted double strand with four point barbs four inches apart is generally used. These fences, like chain link, should also be at least seven feet high and they should, additionally, carry a top guard. Posts should be metal and spaced no more than six feet apart. Vertical distance between strands should be no more than six inches and preferably less.

Concertina Wire

Concertina wire is a coil of steel wire clipped together at intervals to form a cylinder. When it is opened it forms a barrier fifty feet long and three feet high. Developed by the military for rapid laying, it can be used in multiple coils. It can be used either with one roll atop another, or in a pyramid with two rolls along the bottom and one on the top. Ends should be fastened together, and the base wires should be staked to the ground.

Concertina wire is probably the most difficult fence to penetrate, but it is unsightly and is rarely used except in a temporary application.

Walls

In some instances, masonry walls may be used to form all or a part of the perimeter barrier. Such walls may be constructed for aesthetic reasons to replace the less decorative chain link look, or possibly to conceal the operations within that part of the facility.

In those areas where a masonry wall is used, it should be at least seven feet high with a top guard of three or four strands of barbed wire, as in the case of a chain link fence.

Since concealment of the inside activity must be paid for by also cutting off the view of any activity outside the wall, extra efforts must be made to prevent scaling the wall. Ideally, the perimeter line should also be staggered in a way that permits observation of the area in front of the wall from a position or positions inside the perimeter.

Gates and Other Barrier Breaches

Every opening in the perimeter barrier is a potential security hazard. The more gates, the more security personnel must be deployed to supervise the traffic through them. Obviously these openings must be kept to a minimum.

During shift changes there may be more gates open and in use than at other times of the day when the smallest practical number of such gates are in operation. Certainly there must be enough gates in use at any time to facilitate the efficient movement of necessary traffic. The number can only be determined by a careful analysis of the needs at various times of the day—but every effort must be made to reduce the number of operating gates to the minimum consistent with safety and efficiency.

If it develops that some gates are not necessary to the operation of the facility or that they can be dispensed with by changing traffic patterns, the openings should be sealed off and retired from use.

Gates used only at peak periods or emergency gates should be padlocked and secured. If possible, the lock used should be distinctive and immediately identifiable, since it is common for thieves to cut off the plant padlock and substitute their own so they can work at collecting their loot without an alarm being given by a passing patrol that has spotted an otherwise missing lock. A lock of distinctive color or design could compromise this ploy.

It is important that all locked gates be checked frequently. This is especially important where, as is usually the case, these gates are out of the current traffic pattern and are therefore remote from the general activity of the facility.

Personnel gates are usually from four to seven feet wide to permit single line entrance or exit. It is important that they not be so wide that control of personnel is lost. Vehicular gates, on the other hand, must be wide enough to handle the type of traffic typical of the facility. They may handle two-way traffic or, if the need for control is particularly pressing, they may be limited to one-way traffic at any given time.

A drop or railroad crossing type of barrier is normally used to cut off traffic in either direction when the need arises. The gate itself might be single or double swing, rolling or overhead. It could be a manual or an electrical operation. Railroad gates should be secured in the same manner as other gates on the perimeter except during those times when cars are being hauled through it. At these times the operation should be under inspection by a security guard.

Virtually every facility has a number of miscellaneous openings that penetrate the perimeter. All too frequently these are overlooked in security planning, but they *must* be taken into account, because they are frequently the most effective ways of gaining an entrance into the facility without being observed.

These openings or barrier breaches consist of sewers, culverts, drainpipes, utility tunnels, exhaust conduits, air intake pipes, manhole covers, coal chutes and sidewalk elevators. All must be accounted for in the security plan.

Any one of these openings having a cross section area of ninety-six square inches or more must be protected by bars, grillwork, barbed wire, or doors with adequate locking devices. Sidewalk elevators and manhole covers must be secured from below to prevent their unauthorized use. Storm sewers must be fitted with deterrents which can be removed for inspection of the sewer after a rain.

Buildings

When the building forms part of the perimeter barrier or when, in some urban situations, the building walls are the entire perimeter of the facility, it should be viewed in the same light as the rest of the barrier or it should be evaluated in the same way as the outer structural barrier. It must be evaluated in terms of its strength and all openings must be properly secured.

In cases where a fence joins the building as a continuation of the perimeter, there should be no more than two inches between these two structures. Depending on the placement of windows, ledges or setbacks, it might be wise to gradually double the fence height to the point where it joins the building. In such a case the higher section of the fence should extend six to eight feet out from the building.

Windows

Windows and other openings larger than ninety-six square inches should be protected by grilles, metal bars, or heavy screening when they are less than eighteen feet from the ground or when they are less than fourteen feet from structures outside the barrier.

Doors

Doors which penetrate the perimeter walls must be of heavy construction and fitted with a strong lock.

Since both the law and good sense require that there be adequate emergency exits in the event of fire or other danger, provision must be made for such eventualities. Doors that have been created for emergency purposes only should have exterior hardware removed so that they cannot be opened from outside. They can be secured by a remotely operated electromagnetic holding device or they can be alarmed so that their use from inside will be substantially reduced or eliminated.

Roofs and Common Walls

An important, though often overlooked, part of the perimeter is the roof of the building. In urban shopping centers or even in small, free-standing commercial situations where the building walls are the perimeter, entry through the roof is common. Entry can be made through skylights or by chopping through the roof—an activity rarely detected by passers-by or even by patrols.

Buildings sharing a common wall have also frequently been entered by breaking through the wall from a poorly secured neighboring occupancy. All of these means of entry circumvent normal perimeter alarm systems and can, therefore, be particularly damaging.

Barrier Protection

In order for the barrier to be most effective in preventing intrusion, it must be patrolled and inspected regularly. Any fence or wall can be scaled, and unless these barriers are kept under observation, they will be, neutralizing the security effectiveness of the structure.

A clear zone should be maintained on both sides of the barrier to make immediately visible any approach to the barrier from the outside or any movement from the barrier to areas inside the perimeter. Anything outside the barrier, such as refuse piles, weed patches, heavy undergrowth or anything else that might conceal a man's approach, should be eliminated. Inside the perimeter everything should be cleared away from the barrier to create as wide a clear zone as possible.

Unfortunately, it is frequently impossible to achieve an uninterrupted clear zone. Most perimeter barriers are indeed on the perimeter of the property line, which means that there is no opportunity to control the area outside the barrier. The size of the facility and the amount of space needed for its operation will determine how much space can be given up to the creation of a clear zone inside the barrier. It is important, however, to create some kind of a clear zone, however small it must be.

In situations where the clear zone is necessarily so small as to endanger the effectiveness of the barrier, thought should be given to increasing the barrier height in critical areas or to the installation of an intrusion detection device to give due and timely warning of an intrusion to an alert guard force.

Inspection

Having established the perimeter defense and the clear zones to the maximum possible and practical, it is essential that a regular inspection routine be set up. Gates should be examined carefully to determine whether locks or hinges have been tampered with; fence lines should be observed for any signs of forced entry or tunneling; walls should be checked for marks that might indicate they have been scaled or that such an attempt has been made; top guards must be examined for their effectiveness; miscellaneous service penetrations must be examined for any signs of attack; brush and weeds must be cleared away; erosion areas must be filled in; obviously any potential scaling devices such as ladders, ropes, oildrums or stacks of pallets must be cleared out of the area. Any condition which could, even in the smallest degree, compromise the integrity of the perimeter must be reported and corrected.

Such an inspection should be undertaken no less than weekly and possibly more often if the conditions indicate.

Inside the Perimeter

Unroofed or outside areas within the perimeter must be considered a second line of defense, since these areas can usually be observed from the outside and targets selected before an assault is made.

In an area where materials and equipment are stored in a helter-skelter manner, it is difficult for guards to determine if they have been disturbed in any way. On the other hand, storage which is neat, uniform and symmetrical can be readily observed and any disarray can be detected at a glance.

Discarded machinery, scrap lumber and junk of all kinds haphazardly thrown about the area create safety hazards as well as providing cover for any intruder. Such conditions must never be permitted to develop. Efficient housekeeping is basic security.

Parking

The parking of privately owned vehicles within the perimeter barrier should never be allowed. There should be no exceptions to this rule. Facilities which can't or won't establish parking lots outside the perimeter barrier are almost invariably plagued by a high incidence of pilferage, due to the ease with which employees can conceal goods in their cars at any point during the day.

In those cases where the perimeter barrier encompasses the employee and visitor parking area, additional fencing should be constructed to create a new barrier which excludes the parking area. Appropriate guarded pedestrian gates must, of course, be installed to accommodate the movement of employees to and from their cars.

The parking lot itself should be fenced and patrolled to protect against car thieves and vandals. Few things are more damaging to morale than the insecurity which an unprotected parking area in a crime-ridden area can create.

Company cars and trucks—especially loaded or partially loaded vehicles—should be parked within the perimeter for added security. This inside parking area should be well lighted, and it should be regularly patrolled or kept under constant surveillance.

Loaded or partially loaded trucks and trailers should be sealed or padlocked and should, further, be parked close enough together and close enough to a building wall or even back to back so that neither their side doors nor rear doors can be opened without actually moving the vehicles.

Surveillance

The entire outside area within the security barrier must be kept under surveillance at all times, particularly at night. Since goods stored in this area are particularly vulnerable to theft or pilferage, this is the area most likely to attract the thief's first attention. With planning and study in cooperation with production personnel, it will undoubtedly be possible to lay out this yard area so that there are long, uninterrupted sight lines which will permit inspection of the entire area with a minimum of movement.

Lighting

Depending on the nature of the facility, protective lighting will be designed either to emphasize the illumination of the perimeter barrier and the outside approaches to it, or to concentrate on the area and the buildings within the perimeter. In either case it must produce sufficient light to create a psychological deterrent to intrusion as well as to make detection virtually certain in the event an entry is made.

It must avoid glare that would reduce the visibility of security personnel, while creating glare to deter intruders. It must also avoid casting annoying or dangerous lights into neighboring areas. This is particularly important where the facility abuts streets, highways, or navigable waterways.

The system must be reliable and designed with overlapping illumination to avoid creating unprotected areas in the event of individual light failures. It must be easy to maintain and service and it must itself be secured against attack. Poles should be within the barrier, power lines should be buried and the switch box or boxes must be secure.

There should be a back-up power supply in the event of power failure. Supplementary lighting including searchlights and portable lights should also be a part of the system. These lights are provided for special or emergency situations and, although they may not be used with any regularity, they must be available to the security force.

The system could be operated automatically by a photoelectric cell which responds to the amount of light to which it is subjected. Such an arrangement allows for lights to be turned on at darkness and extinguished at daylight. This can be set up to activate individual lamps or to turn on the entire system at one time.

Other controls are timed, which simply means that lights are switched on and off by the clock. Such a system must be adjusted regularly to coincide with the changing hours of sunset and sunrise. The lights may also be operated manually.

Types of Lighting

Lamps used in protective lighting are either incandescent, gaseous discharge or quartz. Each type has special characteristics suitable for specific assignments.

• *Incandescent*—These are common light bulbs of the type found in the home. They have the advantage of providing instant illumination when the switch is thrown and are thus the most commonly used in protective lighting systems. Some incandescents are manufactured with interior coatings which reflect the light, and with a built-in lens to focus or diffuse the light. Regular high-wattage incandescents can be enclosed in a fixture, giving much the same result.

• *Gaseous Discharge Lamps*—Mercury vapor lamps give out a strong light with a bluish cast. They are more efficient than incandescents because of a considerably longer lamp life. Sodium vapor lights give out a soft yellow light and are even more efficient than mercury vapor. They are widely used in areas where fog is a frequent problem, since the yellow light penetrates the mist more readily than a white light. They are frequently found on highways and bridges.

The use of gaseous discharge lamps in protective lighting is somewhat limited, since they require a period of from two to five minutes to light when they are cold and an even longer period to relight, when hot, after a power interruption.

• *Quartz lamps*—These lamps emit a very bright white light and snap on almost as rapidly as the incandescent bulb. They are frequently used at very high wattage—1,500 to 2,000 watts is not uncommon in protective systems—and they are excellent for use along the perimeter barrier and in troublesome areas.

Types of Equipment

No one type of lighting unit is applicable to every need in a protective lighting system, although manufacturers are continually working to develop just such a fixture.

Amid the great profusion of equipment in the market there are four basic types which are in general use in security applications: floodlights, searchlights, Fresnels, and street lights. (The first three of these might, in the strictest sense, be considered as a single type, since they are all basically reflection units in which a parabolic mirror directs the light in various ways. We will, however, deal with them separately.)

Street lights are pendant lighting units which are built as either symmetrical or asymmetrical. The symmetrical units distribute light evenly. Those units are used where a large area is to be lighted without the need for highlighting particular spots. They are normally centrally located in the area to be illuminated.

Asymmetrical units direct the light by reflection in the direction where the light is required. They are used in situations where the lamp must be placed some distance from the target area. Since these are not highly focused units, they do not create a glare problem.

Street lights are rated by wattage or, even more frequently, by lumens, and in protective lighting applications may vary from 4,000 to 10,000 lumens, depending on their use.

Floodlights are fabricated to form a beam so that light can be concentrated and directed to specific areas. They can create considerable glare.

Although many floods specify beam width in degrees, they are generally referred to as wide, medium, or narrow, and the lamp is described in wattage. Lamps may run from 300 to 1,000 watts in most protective applications, but there is a wide latitude in this and the choice of one will depend upon a study of its mission.

Fresnel lights are wide beam units, primarily used to extend the illumination in long, horizontal strips to protect the approaches to the perimeter barrier. Unlike floodlights and searchlights, which project a focused round beam, Fresnels project a narrow, horizontal beam which is approximately 180° in the horizontal and from 15 to 30° in the vertical plane.

These units are especially good for creating a glare for the intruder while the facility remains in comparative darkness. They are normally equipped with a 300 to 500 watt lamp.

Searchlights are highly focused incandescent lamps which are used to pinpoint potential trouble spots. They can be directed to any location within or without the property and, although they can be automated, they are normally controlled manually.

They are rated according to wattage, which may range from 250 to 3,000 watts, and the diameter of the reflector, which may range from six inches to two feet (the average is around eighteen inches). The beam width is from three to ten degrees, although this may vary in adjustable or focusing models.

Maintenance

As with every other element of a security system, electrical circuits and fixtures must be inspected regularly to replace worn parts, verify connections, repair worn insulation, check for corrosion in weatherproof fixtures, and clean reflecting surfaces and lenses. Lamps should be logged as to their operational hours and replaced at between 80 and 90 per cent of their rated life.

Perimeter Lighting

Every effort should be made to locate lighting units far enough inside the fence and high enough to illuminate areas both inside and outside the boundary. The farther outside the boundary the lighted areas extend, the more readily guards will be able to detect the approach of an intruder.

This light should be directed down and away from the protected area. The location of light units should be such that they avoid throwing a glare in the eyes of the guard, they do not create shadow areas, and they do create a glare problem for anyone approaching the boundary.

Fixtures used in barrier and approach lighting should be located inside the barrier. As a rule of thumb, they should be around 30 feet within the perimeter, spaced 150 feet apart, and about 30 feet high. These figures are, of course, approximations and will not apply to every installation. Local conditions will always dictate placement.

Floodlights or Fresnel units are indicated in illuminating isolated or semi-isolated fence boundaries where some glare is called for. In either case, it is important to light from 20 feet inside the fence to as far into the approach as is practical. In the case of the isolated fence, this could be as much as 250 feet. Semi-isolated and non-isolated fence lines cannot be lighted as far into the approach, since such lighting is restricted by streets, highways, and other occupancies.

Since glare cannot be employed in illuminating a non-isolated fence line, street lights are recommended.

Where a building of the facility is near the perimeter or is itself part of the perimeter, lights can be mounted directly on it. Doorways of such buildings should be individually lighted to eliminate shadows cast by other illumination.

In areas where the property line is on a body of water, lighting should be designed to eliminate shaded areas on or near the water or along the shore line. This is especially true for piers and docks, where both land and water approaches must be lighted or capable of being lighted on demand. Before finalizing any plans for protective lighting in the vicinity of navigable waters, however, the United States Coast Guard must be consulted.

Gates and Thoroughfares

It is important that the lighting at all gates and along all interior thoroughfares be sufficient for the operation of the facility.

Since both pedestrian and vehicular gates are normally manned by guards inspecting credentials, as well as inspecting for contraband or stolen property, it is critical that the areas be lighted to at least one footcandle. Pedestrian gates should be lighted to about twenty-five feet on either side of the gate, if possible, and the range for vehicular gates should be twice that. Streetlighting is recommended in these applications, but floodlights can also be used if glare is strictly controlled.

Thoroughfares used for pedestrians, vehicles, or forklifts should be lighted to 0.10 footcandles for security purposes. Much more light may be required for operational efficiency, but this level should be maintained as a minimum, no matter what the conditions of traffic may be.

Other Areas

Open or unroofed areas within the perimeter, but not directly connected to it, require an overall intensity of illumination of about 0.05 footcandles (up to 0.10 footcandles in areas of higher sensitivity). These areas, when non-operational, are usually used for material storage or for parking. Any particularly vulnerable installations in the area should not be lighted at all, but the approaches to them should be well lighted for at least twenty feet to aid in the observation of any movement.

Searchlights may be indicated in some facilities, especially in remote mountainous areas or in waterfront locations where small boats could readily approach the facility.

General

In sum, a well-thought-out plan of lighting along the security barrier and the approaches to it; an adequate overall level of light in storage, parking, and other non-operational areas within the perimeter; and reasonable lighting along all thoroughfares are essential to any basic security program. The lighting required in operational areas will usually be much higher than the minimums required for security and will therefore serve a security purpose as well.

INTERIOR SECURITY

Within any building, whether or not it is located inside a perimeter barrier other than the walls of the building itself, it is necessary to consider the need to protect against the internal thief as well as the intruder. Whereas the boundary fence is primarily designed to keep out unwanted visitors (not altogether forgetting its function in the control of movement of authorized personnel), interior security must provide some protection against the free movement of employees bent on pilferage as well as establishing a second line of defense against the intruder.

Since every building is used differently and has its own unique traffic composition and flow, each building presents a different security problem. Each must be examined and analyzed in great detail before a really effective security program can be developed.

It cannot be overemphasized that such a program must be implemented without in any way interfering with the orderly and efficient operation of the facility to be protected. It must not be obtrusive and yet it must provide a predetermined level of protection against criminal attack from without or within.

The first points of examination must be the doors and windows. These must be considered in terms of effectiveness whether the building walls form a part of or constitute in themselves the perimeter barrier, as we have already discussed, or whether they are a true second defense line where the building under examination is completely within the protection of a barrier.

Windows

It is axiomatic that windows should be protected. Since the ease with which most windows can be entered makes them a ready target for the intruder, they must be viewed as a potential weak spot in any building's defense.

In most industrial facilities, windows should be protected with grillwork, heavy screening, or chain link fencing. In cases where caution dictates they may be needed as emergency exits beyond strict requirements of fire laws, or where they might be needed to

lead in fire hoses, consideration should be given to hinging and pad-
locking with protective coverings.

Burglary-Resistant Glass

In applications such as prominent administration buildings, office
buildings, and the like, where architectural considerations preclude
the use of such relatively clumsy installations, the windows can be
immeasurably strengthened by the use of either UL-listed (a term
which means that the material or item so designated has met certain standards of Underwriters' Laboratories for burglary resistance) burglary-resistant glass, or one of several brands of UL-listed
polycarbonate or acrylic glazing materials. Both of these products
are considerably more expensive than plate glass and are generally
used only in those areas where attack can be expected or where a
reduction in insurance premium would justify the added expense.

As opposed to tempered glass, which is designed to protect
people from the danger of flying shards in the event of breakage,
UL-listed burglary-resistant glass resists heat, flame, cold, hammers,
picks, rocks, and virtually anything an intruder might use. For all
practical purposes it is impenetrable.

The plastic glazing materials are equally resistant, although some
of them are subject to scratches and defacement under an attack.
One major manufacturer of plastic glazing materials advertises an
impact strength 300 times greater than tempered glass.

Both materials have the appearance of ordinary glass. Obviously
any window so hardened against entry must be securely locked from
the inside to protect against intrusion from the outside.

"Smash and Grab" Attacks

Burglary-resistant glass is used to a considerable degree in banks
and retail stores where there has been a very real need to prevent
the "smash and grab" raid on window displays and showcases.

It should be noted that UL-listed burglary-resistant glass is a
laminate of two sheets of flat glass (usually 3/16" thick) held
together by a 1/16" layer of polyvinyl butyral, a soft transparent
material. In this thickness, laminated glass is virtually indistinguishable from ordinary glass; hence the burglar sets about with a
hammer or iron bar to break what he supposes to be a plate glass
window. It is only after he has made a few unsuccessful tries that

he realizes he is up against a material he is not apt to penetrate, certainly not in the time he has between first attack and police or guard response.

Even though the attacker may flee empty-handed, the owner in such situations is left with a window with a web of cracks over the surface of the outer layer of glass, making replacement of the entire pane necessary in applications where appearance is important. Insurers are, in many cases, requiring that laminated glass be clearly identified so as to discourage what would be a futile but damaging assault by the "smash and grab" attacker.

Screening

It can also be important to screen windows to protect against their use as a means by which employees can temporarily dispose of goods for later recovery. The smaller the goods being manufactured or available on the premises, the smaller the mesh in the screen must be to protect against this kind of pilferage.

Generally speaking, any windows less than eighteen feet from the ground or less than fourteen feet from trees, poles, or adjoining buildings should receive some protective treatment unless they are well within the perimeter barrier and open directly onto an area outside the building which is particularly well secured.

Doors

Every door within the building must be carefully examined to determine the degree of security required of it. Such an examination will also determine the type of construction as well as the locking system to be used on each door.

Whatever security measures may be required at any specific door will be determined by the operations in progress or by the value of the assets stored or available in the various areas. The need for adequate security cannot be overemphasized, but it must be provided as part of an overall plan for the safe and efficient conduct of the business at hand.

When this balance is lost, the business must suffer. Either the security function will be downgraded in favor of a more immediate convenience, or the smooth flow of business will be impeded to conform to obtrusive security standards. Either of these conditions is intolerable in any business, and it is a management responsibility to determine the balance required in establishing systems that will recognize and accommodate all such company needs.

Door Construction and Hardware

Doors are frequently much weaker than the surface into which they are positioned. Panels may be thin, easily broken wood, or glass. Locks may be old and ineffective. The door frame may be so constructed that a lever or a plastic card can be inserted between the door and jamb to disengage the bolt in the lock. Even with a properly hung door, if the jamb is of soft, unreinforced aluminum it can be peeled or ripped away from a long-throw bolt. Heavy wood or metal doors with reinforced jambs can go a long way to overcome these problems.

In some cases, doors are entered by "pulling"—a technique whereby the lock cylinder is ripped from the door and the lock is operated through the opening left in its face. The installation of a special, hardened-steel cylinder guard can overcome this kind of assault.

Door hinges may also contribute to a door's weakness. Surface mounted hinges with mounting screws or hinge pins exposed on the exterior side of the door can be removed; thus entrance can be gained on the hinge side. To complicate the matter, the door can be replaced on its hinges after the intruder has finished his business and, in most cases, the intrusion would never be detected. Without any visible sign of forced entry, very few insurance policies would pay off on the stolen merchandise.

To prevent this unhappy chain of events, hinges should be installed with the screws concealed and with the hinge pins either welded or flanged to prevent removal.

Attack by Picking

Although direct forcible assault is the method generally used to gain entry, more highly skilled burglars may concentrate on the lock. This may be their only practical means of ingress if the door and jamb are well-designed in security terms and essentially impervious to forcible attack.

Picking the lock or making a key by impression are the methods generally used; both require a degree of expertise. In the former method, metal picks are used to align the levers or tumblers as an authorized key would, thus enabling the lock to operate. Making a key by taking impressions is a technique requiring even greater skill, since it is a delicate, painstaking operation requiring repeated trials.

Because both of these techniques are apt to take time, they are customarily used to attack those doors where the intruder feels he may work undisturbed and unobserved for adequate periods of time. Here again—the picked lock rarely shows any signs of illegal entry, and often the insurance is uncollectible.

Pick-Resistant Locks

The best defense against these methods is the installation of special pick-resistant, impression-resistant lock cylinders. They are more expensive than standard cylinders but in many applications may well be worth the added cost. (Regular patrols or constant CCTV—closed circuit television—surveillance as part of the overall security system are also effective deterrents to any such intrusion.)

Although these special cylinders are highly effective and can be very important in equipping a facility's security defense, they are only one part of the whole. Locks, cylinders, door and frame construction, and key control are inseparable elements of door and entrance security; all must be equally effective. If any one element is weak, the system breaks down.

Key Control

A careful, strictly supervised record of all keys issued must be maintained by the security department. This record should indicate the name and department of the person to whom the key was issued as well as the date of issue.

A key depository for securing keys during non-working hours should be centrally located, locked, and kept under the supervision of security personnel. Keys issued on a daily basis, or those issued for a specific one-time purpose, should be accounted for daily. Keys should be counted and signed for by the security supervisor at the beginning of each working day.

When a key is lost, the circumstances should be investigated and set forth in writing. In some instances, if the lost key provides access to sensitive areas, locks should be changed.

Master keys should be kept to a minimum. If possible, submasters should be used, and they should be issued only to a limited list of personnel especially selected by management. This list should be reviewed periodically to determine whether all those authorized should continue to hold such keys.

Before a decision can be reached with respect to the master and sub-master key systems and how such keys should be issued, there must be a careful survey of existing and proposed security plans, along with a study of current and planned locking devices. Where security plans have been developed with operational needs of the facility in mind, the composition of the various keying systems can be readily developed.

Keying Systems

The three types of keys used in such systems are:

· *The change key*—a key to a single lock within a master-keyed system.

· *The sub-master key*—a key that will open all the locks within a particular area or grouping in a given facility. In an office, a sub-master might open all doors in the Accounting Department; in an industrial facility, it might open all locks in the loading dock area. Normally such groupings concern themselves with a common function or they may simply be located in the same area, even if they are not otherwise related.

· *The master key*—where two or more sub-master systems exist, a master key system is established. Such a key would open any of the systems.

Obviously, master and sub-master keys must be treated with the greatest care. If a master key is lost, the entire system is threatened. Rekeying is the only really secure thing that can be considered, but the cost of such an effort can be enormous.

Any master key system is vulnerable. Beyond the danger of loss of the master itself and the subsequent staggering cost of rekeying— or, even more unfortunate, the use of such a key by enterprising criminals to loot the facility—there is the problem that it necessarily serves a lesser lock. Locks in such a system are neither pick-resistant nor resistant to making a key by impression.

On the other hand, relative security coupled with convenience may make such a system preferable in some applications where it would not be in others. Only the most careful evaluation of the particular circumstances of a given facility will determine the most efficient and most effective keying system.

In any event, the keys must be strictly and methodically controlled. Every effort should be exerted to develop ways whereby keys remain in the hands of security personnel or management personnel; and, in those cases where this is not possible or practical, there must be a system of inventory and accountability.

For example, keys should never be issued on a long-term basis to outside janitorial personnel. The high employee turnover rate in this field would suggest that this could be a dangerous practice. Employees of this service should be admitted by guards or other building employees and issued interior keys which they must return before leaving the building.

By the same token, it is bad practice to issue entrance keys to tenants of an office building. If such is done, control of this vital security point is lost. A guard or building employee should control entry and exit before and after regular building hours. If keys must be issued to tenants, however, the lock cylinder in the entrance should be changed every few months and new keys issued to authorized tenants.

Traffic Patterns

Doors must be analyzed for their function in laying out the security plan. In some cases they may serve a dual purpose, as, for example, fire doors which are designed to close automatically in the event of a fire. These doors, which may remain open at the discretion of management, must be fitted to form an effective and automatic barrier to the spread of fire, but they may also be kept closed if that is suitable to the circumstances. This may be desirable when fire doors separate a production area from a warehouse or storage area. During those times when the production area is in operation but the warehouse is not, such fire doors can perform a security function by remaining closed.

In other cases, doors must be examined in an effort to establish a schedule for their use. Employee entrances that are the authorized points of passage for all employees may be manned by security personnel, depending on whether the control point is established there or farther out on the perimeter. These doors could be secured once the employees have entered, thus denying entrance to unauthorized visitors as well as preventing any employees from wandering out to the fence or the parking lot or any other location where they might cache contraband for later pick-up or transport.

It is axiomatic, however, that any door used as an entrance will, in a time of emergency, be used as an exit by some employees. This is true in apartments and office buildings, and even in industrial facilities where the employees are thoroughly familiar with the premises. No matter what or how many designated emergency exits or procedures there may be, some individuals in a time of tension or near-panic will seek out the door with which they are most familiar.

The entrance, then, must always be considered an emergency exit, and it should be equipped with panic hardware. To protect against surreptitious use it should also be fitted with a local alarm.

The same, of course, is true of the designated emergency exits. These doors should additionally be stripped of all exterior hardware from the outside, since they are not intended for operational use at any time.

Personnel doors leading to and from the dock area must be carefully controlled and supervised at all times when the dock is in use; these and dock doors must be secured when the area is no longer operational.

Fire doors in office buildings should be alarmed to prevent surreptitious use, and access to interior, public stairwells should be prohibited or discouraged unless doors from them open out into reception areas.

Doors to Sensitive Areas

Doors to telephone equipment rooms, computer installations, R & D and other sensitive areas should be equipped with automatic door closing devices and fitted with a strong dead bolt and a heavy latch.

In cases where an area is under heavy security, but has any degree of traffic, it might be well to consider the installation of an electric strike to secure the operation and control the traffic. This kind of unit is a locking device controlled remotely by a security person, permitting entry of a recognized, authorized person only when a button is pressed to release the lock. Since it requires someone on hand at all times for its operation, this system can be expensive; it must be examined with this cost-vs.-security-cost equation in mind.

Supply rooms and toolroom doors should be secured whenever

these rooms are not actually in operation. Even when they are, entrance into these areas must be restricted. The usual construction of such restraints consists of either a dutch door in which the bottom half is secured, or a counter which can be closed off by heavy screening, chain link fencing material, or reinforced shutters.

Special care should be taken in the storage of small items of value. Such merchandise or material is highly pilferable by virtue of its value for resale or personal use combined with the ease with which it can be stolen. Although such items may be stored in a facility of any construction capable of providing security, it has been the experience of many firms that uniformly stacked rows, piles, or pallets of such items within a cage-type construction that provides instant eyeball inventory is the best protection. Such precautions will vary from business to business, but they must be carefully systematized to control this potentially troublesome area of loss.

Office Area Doors

Doors between production and office areas, or heavily trafficked areas and office spaces, must be examined for the likelihood of their use for criminal purposes. Their construction and locking hardware will be determined by such a survey. In most cases these passages would be minimum security areas during regular working hours, since there is usually a need for movement between these areas. When there is little or no use of the office area, these doors should be secured.

Locking Schedules

Door-locking schedules and responsibilities must be established and supervised vigorously. The system must be set up in such a way that a procedure for altering the routine to fit immediate needs is possible, but in all respects the schedule, whether the master plan or the temporary, must be adhered to in every detail. A breakdown in such a system, especially in large offices, institutions, or industrial facilities, could represent just the opportunity that an alert criminal is waiting for.

Old Construction

Older buildings, particularly though certainly not exclusively office buildings, present a host of different and difficult problems to the security staff. Exterior fire escapes, old and frequently badly worn locks, common walls, roof access from neighboring buildings, unused and forgotten connecting doors – all increase the exposure to burglary.

It is vital that all such openings be surveyed and plans made for securing them. Those windows not designated as emergency exits must be barred or screened. Where windows lead to a fire escape or are accessible to adjacent fire escapes, their essential security must be accomplished within the regulations of the local fire codes. Fire safety must be a primary consideration. In cases where prudence or the law (or both) dictate that locks would be a hazard to safety, windows should be alarmed and the interior areas to which these windows provide access must be further secured. Here security can be likened to an army retreating to secondary or tertiary lines of defense to establish a strong and defensible position.

It is also well to consider the danger of attack from neighboring occupancies in shared space where entry might be made from a low risk, badly secured premise into a higher risk area that might otherwise be well protected against a more direct attack.

New Construction

Modern urban buildings, though security conscious in varying degrees, present their own problems. Most interior construction is standardized. Fire and building codes are such that corridor doors can resist most attack if the hardware is adequate. Corridor ceilings are fixed, and entrances to individual offices usually offer a fairly high degree of security.

On the other hand, modern construction creates offices which are essentially open top boxes. It has solid exterior walls (though interior walls are frequently plaster board) and a concrete floor. But nothing of any security value protects the top. The ceiling is simply a layer of acoustical tiles lying loose on runners suspended between partition walls. In the space above these tiles—between them and the concrete slab above—are vital air conditioning ducts, and wiring for power and telephones.

In effect, any given floor of a building has a crawlspace that runs from exterior wall to exterior wall. This may not be literally so in every case, but the net result stands. It means that virtually every room and every office is accessible through this space. Once this crawlspace is reached from any occupancy, the remaining offices on that floor are accessible.

Extending dividing walls up to the next floor will not solve the problem, since this "drywall" construction is easily broken through and, in any case, it must be breached to allow passage of all utilities. Alarms of various kinds, which are discussed later, are recommended to overcome this problem.

FILES, SAFES, AND VAULTS

The final line of defense at any facility is in the high security storage areas where papers, records, plans, cash or cashable instruments, precious metals, or other especially valuable assets are protected. These security containers will be of a size and quantity which the nature of the business dictates. (Various characteristics of security containers are discussed in greater detail in Chapter 4.)

Every facility will have its own particular needs, but certain general observations apply. The choice of the proper security container for specific applications is influenced largely by the value and the vulnerability of the items to be stored in them. Irreplaceable papers or original documents may not have any intrinsic or marketable value, so they may not be a likely target for a thief; but since they do have great value to the owners, they must be protected against fire. On the other hand, uncut precious stones, or even recorded negotiable papers which can be replaced, may not be in danger from fire, but they would surely be attractive to a thief; they must therefore be protected from him.

In protecting property, it is essential to recognize that, generally speaking, protective containers are designed to secure against burglary or fire. Each type of equipment has a specialized function, and each type provides only minimal protection against the other risk. There are containers designed with a burglary-resistant chest within a fire-resistant container which are useful in many instances, but these, too, must be evaluated in terms of the mission.

Whatever the equipment, the staff must be educated and reminded of the different roles played by the two types of container. It is all too common for company personnel to assume that the fire-resistive safe is also burglary-resistant, and vice versa.

Files

Burglary-resistant files are secure against most surreptitious attack. On the other hand, they can be pried open in less than half an hour if the burglar is permitted to work undisturbed and is not concerned with the noise created in the operation. Such files are suitable for non-negotiable papers or even proprietary information, since these items are normally only targeted by surreptitious assault.

Filing cabinets with fire-rating of one hour, and further fitted with a combination lock, would probably be suitable for all uses but the storage of government classified documents.[4]

Safes

Burglary-resistive safes (Class E through H) are secure in increasing degrees from most successful attacks. The particular type should always be chosen to protect the maximum amount that it is likely to hold.

Since burglary-resistant safes have a limited holding capacity, it is always advisable to study the volume of the items to be secured. If the volume is sufficiently large, it might be advisable to consider the installation of a burglary-resistant vault which, although considerably more expensive, can have an enormous holding capacity.

Whatever safe is selected must be securely fastened to the structure where it is located. Police reports are filled with cases where unattached safes, some as heavy as a ton, have been stolen in their entirety—safe and contents—to be worked on in uninterrupted concentration.

A convicted criminal recently told investigators how he and an accomplice had watched a supermarket to determine the cash flow and the manager's banking habits. They noted that he built up cash in a small wheeled safe until Saturday morning, when he banked. Presumably he felt secure in this practice, since he lived in an apartment above the store and perhaps felt that he was very much on top of the situation in every way. One Friday night the thief and his friend rolled the safe into their station wagon; they pried it open at their leisure to get the $15,000 inside.

Pleased with their success, the thieves were even more pleased when they found that the manager replaced the stolen safe with one exactly like it and continued with the same banking routine. Two weeks later, our man went back alone and picked up another $12,000 in exactly the same way as before.

It is becoming a common practice to install the safe in a concrete floor where it offers great resistance to attack. In this kind of installation, only the door and its combination are exposed. Since the door is the strongest part of a modern safe, the chances of successful robbery are considerably reduced.

Vaults

Vaults are, essentially, enlarged safes. As such they are subject to the same kinds of attack and must look at the same basic principles of protection as safes.

Since it would be prohibitively expensive to build a vault out of shaped and welded steel and special alloys, the construction, except for the door, is usually of high quality reinforced concrete. There are many ways in which such a vault can be constructed, but however it is done it will always be extremely heavy and, at best, a difficult architectural problem.

Normally vaults are situated at or below ground level so they do not add to the stresses of the structure housing them. If a vault must be built on the upper stories of a building, it must be supported by independent members which do not provide support for other parts of the building. And it must be strong enough to withstand the weight imposed upon it if the building should collapse from under it as the result of fire or explosion.

The doors of such vaults are normally 6" thick, and they may be as much as 24" in the largest installations. Since these doors present a formidable obstacle to any criminal, an attack will usually be directed at the walls, ceiling, or floor, which must match the strength of the door. As a rule, these surfaces should be twice as thick as the door and never less than 12".

If at all possible, a vault should be surrounded by narrow corridors which will permit inspection of the exterior, but which will be sufficiently confined to discourage the use of heavy drilling or cutting equipment by attackers. It is important that there be no power outlets anywhere in the vicinity of the vault.

Container Protection

Since no container can resist assault indefinitely, it must be supported by alarm systems and frequent inspections. Ideally, any container should be inspected at least once within the period of its rated resistance. CCTV surveillance can, of course, provide constant inspection and, if the expense is warranted, is highly recommended.

By the same token, safes have a greater degree of security if they are well lighted and located where they can be seen readily. Any safe located where it can be seen from a well-policed street will be much less likely to be attacked than one which sits in a darkened back office on the upper floors.

Continuing Evaluation

Security containers are the last line of defense, but in many situations they should be the first choice in establishing a sound security system. These containers must be selected with care after an exhaustive evaluation of the needs of the facility under examination. They must also be reviewed regularly for their suitability to the job they are to perform.

Just as the safe manufacturers are continually improving the design, construction, and materials used in safes, so is the criminal world improving its technology and technique of successful attack. Because of the considerable capital outlay involved in providing the firm with adequate security containers, many businessmen are reluctant to entertain the notion that these containers may someday become outmoded—not because they wear out or cease to function, but because new tools and techniques have nullified their effectiveness.

In selecting security containers it is important that the equipment conform to the needs of the risk and that it be regularly re-evaluated and, if necessary, brought up to date, however unwelcome the additional outlay may be.

TRAFFIC CONTROL

Controlling traffic in and out and within a facility is essential to its security program. Perimeter barriers, locked doors and screened windows prevent or deter the entry of unauthorized visitors, but since some traffic is essential to every operation, no matter how highly classified it may be, provision must be made for the control of this movement.

Specific solutions will depend upon the nature of the business. Obviously retail establishments, which encourage high volume traffic and which regularly handle a great deal of merchandise both in and out, have a problem of a different dimension from the industrial operation working on a highly classified government project. Both, however, must work from the same general principles toward providing the greatest possible security within the efficient and effective operation of the job at hand.

Controlling traffic includes the identification of employees and visitors and directing or limiting their movement; the control of all incoming and outgoing packages; and control of trucks and private cars.

Visitors

All visitors to any facility should be required to identify themselves. When allowed to enter after establishing themselves as being on an authorized call, they should be limited to predetermined, unrestricted areas.

If possible, sales, service, and trade personnel should receive clearance in advance upon making an appointment with the person responsible for their being there. Although this is not always possible, most businesses deal with such visitors on an appointment basis and a system of notifying the security personnel can be established in a majority of cases.

Businesses regularly called upon unannounced by salesmen or other trades people should set aside a waiting room which can be reached without passing through sensitive areas. In some cases, it may be advisable to issue them a pass which clearly designates them as visitors. If they will be escorted to and from their destination, a pass system is probably unnecessary.

Ideally, all traffic patterns involving visitors should be short, physically confined to keep them from straying, and capable of being observed at all points along the route. In spread-out industrial facilities, they should take the shortest, most direct route that will not pass through restricted, sensitive, or dangerous areas, and will pass from one reception area to another.

To achieve security objectives without alienating visitors and without in any way interfering with the operation of the business, any effective control system must be simple and understandable. It must incorporate certain specific elements in order to accomplish its aim. It must limit entry to those people who are authorized to be there, and it must be able to identify such people. It must have a procedure by which persons may be identified as being authorized to be in certain areas. It must prevent theft, pilferage, or damage to assets of the installation. And it must prevent injury to the visitor.

Employee Identification

Small industrial facilities and most offices find that personal identification of employees by guards or receptionists is adequate protection against intruders entering under the guise of employees. In plants of over fifty employees per shift, or in high turnover businesses, this type of identification is inadequate. The opportunity for error is simply too great.

Badges

The most practical and generally accepted system is the use of badges or identification cards. Generally speaking, this system should designate when and where and how passes should be displayed, and to whom; what is to be done in case of the loss of the pass; procedures for taking badges from terminating employees; and a system for a cancellation and re-issue of all passes, either as a security review or when a significant number of badges have been reported lost or stolen.

To be effective, badges must be tamper-resistant, which means that they should be printed or embossed on a distinctive stock which is worked with a series of designs difficult to reproduce. They should contain a clear and recent photograph of the bearer, preferably in color. The photograph should be at least one inch square and should be updated every two or three years or when there is any significant change in facial appearance, such as the growing or removal of a beard or moustache. It should additionally contain vital statistics, such as date of birth, height, weight, color of hair and eyes, sex and both thumbprints. It should be laminated and of sturdy construction. In cases where there are areas set off or restricted to general employee traffic, it might be color coded to indicate those areas to which the bearer has authorized access.

If a badge system is established it will only be as effective as its enforcement. Facility guards are responsible to see that the system is adhered to, but they must have the cooperation of the majority of the employees and the full support of management. If the system is simply a *pro forma* exercise, it becomes a useless annoyance and could better be dispensed with.

Package Control

Every facility must establish a system for the control of packages entering or leaving the premises. However desirable it might seem, it is simply unrealistic to suppose that a blanket rule forbidding packages either in or out would be workable. Such a rule would be damaging to employee morale and, in many cases, would actually work against the efficient operation of the facility. Therefore, since the transporting of packages through the portals is a fact of life, they must be dealt with in order to prevent theft and misappropriation of company property.

If it is deemed necessary, the type of items that may be brought in or taken out may be limited. If such is the case, the fact must be publicized and clearly understood by everyone.

Packages brought in should be checked as to content. If possible, where they are not to be used during work, they should be checked with the guard to be picked up at the end of the day. In most cases, spot checking will suffice.

Vehicle Control

Vehicular traffic within the boundaries of any facility must be carefully controlled, for safety as well as to control the transporting of pilfered goods from the premises.

Merchandise and materials can be readily concealed in or under employee cars, company and outside trucks, railroad cars, or any vehicle that has access to the facility. The more readily these vehicles may drive through the area or approach operating zones, the more acute the problem can become.

It is important, therefore, to make every effort to separate the parking area from all other areas of the facility. This area must be protected from intruders, but the parking lot itself must be separated from production, distribution, and storage areas. Employees and visitors going to and from their cars will pass through pedestrian gates where they can be identified by security personnel.

It would be well to make a most careful evaluation of this vital need. Even the high cost of industrial real estate should not prevent the allocation of space for a parking facility which is separated by a barrier from the rest of the premises. Time and again, firms have reported losses disappearing to almost nothing when they have isolated the parking lot. This is an essential feature of any industrial operation today.

INSPECTIONS

In spite of all defensive devices, the possibility of an intrusion always exists. The highest fence can be scaled and the stoutest lock can be compromised. Even highly sophisticated alarm systems can be contravened by a knowledgeable professional. The most efficient system of physical protection can, at some time or another, be foiled.

It is necessary, therefore, to continually support each element of the system with another—to remember the concept of defense in depth. The ultimate backup—surveillance—must never let down.

Guard Patrols

Visual inspections by irregular patrols through office spaces or through an industrial complex, or constant CCTV surveillance of these same areas, are vital to the success of the security program.

It is equally important to "sweep" the facility after closing time. "Hide-ins" are common in offices or retail establishments. These are thieves who conceal themselves in a closet or utility room and wait for the establishment to close and everybody to go home. After the "hide-in" takes what he is looking for, his only challenge is to break out. The chances of catching such a thief in a premise protected only by perimeter alarms is remote indeed. He must be picked up on the "sweep" when guards go through the entire facility from top to bottom or from east to west to see that everyone required to do so has left.

Specific duties of guards on patrol duty are discussed in some detail in the section dealing with guard forces, but in general it should be noted that patrols should be made at least once each hour, and more often if the area and the size of the guard force permit.

Particular attention must be paid to any signs of tampering with locks, gates, fences, doors, or windows. The presence of any rubbish or piles of materials should be noted for the possibility of concealment—particularly if they are near the perimeter barrier or in the vicinity of storage areas.

In the patrol of office buildings, it is wise to stop occasionally for a long enough period of time to listen for any sounds that might indicate the presence of an intruder.

It is equally important that patrols in any facility be alert to any condition that might prove to be hazardous. These might be anything from an oil slick on a normally trafficked area to a heater left on and unattended. Those conditions presenting an immediate danger must be corrected immediately; others must be reported for correction. All of them must be noted in the log and on the appropriate form.

FIRE

There are no fireproof buildings, however frequently the term may be misapplied. There are *fire-resistive* buildings. But since even these are filled with tons of combustible materials such as furnishings, panelling, stored flammable materials and so on and on, they can become an oven that does not itself burn but can generate heat of sufficient intensity to destroy everything inside it. Eventually such heat can even soften the structural steel to such an extent that part or all of the building may collapse. By this time, however, the collapse of the building endangers only outside elements, since everything inside, with the possible exception of certain fire-resistant containers and their contents, will have been destroyed.

The particular danger of this situation is that, while wooden or wood frame construction can be recognized for the fire hazard they represent, many otherwise knowledgeable people are oblivious to the potential dangers from fire in steel and concrete construction. And the danger can be one which is largely unrecognized because it may grow slowly.

The degree of fire exposure in any fire-resistive building is dependent upon its fire-loading—the amount of combustible materials that occupy its interior spaces. In the case of multiple occupancies, such as large office buildings, no one office manager can control the fire-loading—hence the risk, since the safety of anyone's premises is dependent upon the fire load throughout the entire building. In such an environment, new furniture, decorative pieces, drapes, carpeting, unprotected insulated cables, or even volatile fluids for cleaning or lubricating are piled in every day. And the classic triangle of fire grows larger with each such addition.

The Nature of Fire

The classic triangle so frequently referred to in describing the nature of fire consists of heat, fuel, and oxygen. If all three exist there will be fire; remove or reduce any one and the fire will be reduced or extinguished.

Certainly fuel and oxygen are always present. It would be difficult to imagine any facility that had no combustible items exposed, and air must be present. Only sufficient heat is missing, and that is readily supplied by a careless cigarette or faulty wiring, two of the most common causative factors.

There are, in fact, an almost infinite number of heat sources which can complete the deadly triangle and start a fire raging in virtually any facility. Every fire prevention program must start by controlling the amount and nature of the fire load or fuel, and by instituting programs to prevent the occurrence of any heat build-up, whether from careless smoking or sparks from a welding torch.

Fire Prevention

It should be noted here that any defense against fire must be viewed in two parts. Fire prevention, which is usually the major preoccupation, embodies the control of the sources of heat and the elimination or the isolation of the more obviously dangerous fuels. This commendable effort to prevent fire must not, however, be undertaken at the expense of an equal effort for fire *protection.*

Fire Protection

Fire protection includes not only the equipment to control or extinguish fire, but also those devices which will protect the building, its contents, and particularly its occupants, in the event of fire. Fire doors, fire walls, smokeproof towers, fire safes, non-flammable rugs and furnishings, fire detector systems—all are fire protection matters and are essential to any fire safety program.

By-Products of Fire

Contrary to popular opinion, flame or visible fire is rarely the killer in the nearly 12,000 deaths from fire that occur annually in this country. These are usually caused by smoke or heat, or from gas, explosion or panic. Several such by-products accompany every fire; all must be considered when defenses are being planned.

Smoke will blind and asphyxiate—and in an astonishingly short time. Tests have been conducted in which smoke in a corridor reduced the visibility to zero in two minutes from the time of ignition. A stairway *two feet* from a subject in the test was totally obscured.

Gas, which is largely carbon dioxide and carbon monoxide, collects under pressure in pockets in the upper floors of the building. As the heat rises and the pressure increases, explosions can occur.

Heat expands and creates pressure. It ignites more materials, explodes gas.

Expanded air created by the heat creates fantastic pressure, which will shatter doors and windows and travel at crushing force and speed down every corridor and through every duct in the building.

All of these elements move upward and therefore permit some control over the direction of the fire—if the construction of the building has been planned with proper fire measures in mind.

Classes of Fire

All fires are classified in one of four groups. It is important that these groups and their designation be widely known, since the use of various kinds of extinguishers is dependent upon the type of fire to be fought.

• *Class A:* Fires in ordinary combustible materials such as waste paper, rags, drapes, furniture, etc. These fires are most effectively extinguished by water or water fog. It is important to cool the entire mass of burning materials to below the ignition point to prevent rekindling.

• *Class B:* Fires fueled by substances such as gasoline, grease, oil, or volatile fluids. Such fluids are used in many ways and may be present in virtually any facility. Here a smothering effect such as CO_2 is used. A stream of water on such fires would simply serve to spread the substances with disastrous results. Water fog is excellent, since it cools without spreading the fuel.

• *Class C:* Fires in live electrical equipment such as transformers, generators, or electric motors. The extinguishing agent must be non-conductive to avoid danger to the fire fighter.

• *Class D:* Fires involving certain combustible metals, such as magnesium, sodium, potassium, etc. Dry powder is usually the most, and in some cases the only, effective extinguishing agent. Because these fires can occur only where such combustible metals are in use, they are fortunately rare.

Extinguishers

The security department must evaluate the fire risk for each facility and determine the types of fires to which it might be exposed. Every operation is probably potentially subject to Class A and C fires, and most would be additionally threatened by Class B fires to some degree.

Having made such a determination, security must then select the types of fire extinguishers most likely to be needed. The choice of extinguisher is not difficult, but it can only be made after the nature of the risks are determined. Extinguisher manufacturers can supply all pertinent data on the equipment they supply, but the general types in general use should be known.

Types of Extinguishers

Soda and Acid—These water-base extinguishers are effective on small Class A fires. They are, however, heavy and somewhat clumsy, and the stream is small.

Dry Chemical—These were originally designed for Class B and C fires. The new models now in general use are also effective on Class A fires, since the chemicals are flame-interrupting and in some cases act as a coolant.

Dry Powder—Used on Class D Fires. Smothers and coats.

Foam Extinguishers—effective for small Class A and B fires where blanketing is desirable.

CO_2—Generally used on Class B or C fires, they can be useful on Class A fires, though the CO_2 has no lasting cooling effect.

Carbon Tetrachloride—These are still in fairly wide use, though they are no longer recommended in the National Fire Protection Association Extinguisher Standard. They are usually rated for use on Class B and C fires. The liquid carbon tetrachloride vaporizes when exposed to heat and smothers the fire by oxygen exclusion. Great care must be exercised in the use of these extinguishers in closed spaces, however, since the fumes are very toxic and can be dangerous to the operator.

Water Fog—Fog is one of the most effective extinguishing devices known for dealing with Class A and B fires. It can be created by a special nozzle on the hose or by an adjustment of an all-purpose nozzle similar to those found on a garden hose.

Advantages of Fog

It might be useful here to look briefly at the advantages fog has over a solid stream of water.

- It cools the fuel more quickly.
- It uses less water for the same effect, so water damage is reduced.
- Because fog reduces more heat more rapidly, atmospheric temperatures are quickly reduced. Persons trapped beyond a fire can be brought out through fog.
- The rapid cooling draws in fresh air.
- Fog reduces smoke by precipitating out particulate matter as well as by actually driving the smoke away from the fog.

Inspection and Maintenance

After extinguishers have been installed, a regular program of inspection and maintenance must be established. A good policy is for security personnel to visually check all devices once a month, and to have the extinguisher service company inspect them twice a year. In this process the serviceman should re-tag and, if necessary, recharge the extinguishers and replace defective equipment.

Education in Fire Prevention and Safety

Educating employees about fire prevention, fire protection, and evacuation procedures should be a continuous program. Ignorance and carelessness are the causes of most fires and of much loss of life. An ongoing fire safety program will inform all employees and help to keep them aware of the ever-present, very real danger of fire.

Such a program would ideally include evacuation drills. Since such exercises require shutting down operations for a period of time and lead to the loss of expensive man-hours of productive effort, management is frequently cool toward them.

Basic Fire Training

However, indoctrination sessions for new employees and regular review sessions for all personnel are essential. Such sessions should be brief and involve only a small group. They should include the following subjects as well as many others that may have particular application to the specific facility:

- Walk to primary and secondary fire exits and demonstrate how such exits are opened. Emphasize the importance of closing exit stairwells. If possible, employees should walk down these stairs.

- Explain how to report a fire. Emphasize the need to report first before trying to put out the fire.

- Distribute a simple plan of action in the event of fire.

- Explain the alarm system.

- Explain the need to react quickly and emphasize the need to remain calm and avoid panic.

- Explain that elevators are *never* to be used as an emergency exit.

- Point out the danger of opening doors. Explain that doors must be felt before being opened. Opening a hot door is usually fatal.

- Demonstrate available fire fighting equipment or show manufacturers' film or fire department film on similar equipment.

- Describe what should be done if escape is cut off by smoke or fire:

  a) Move as far from fire as possible.
  b) Move into building perimeter area with a solid door.
  c) Remove readily flammable material out of that area, if possible.
  d) Since expanded air exerts enormous pressures, barricade the door with heavy, non-combustible materials.
  e) Open top and bottom of windows. Fire elements exhaust through the top while cool air will enter through the bottom.
  f) Stay near the floor.
  g) Hang something from windows to attract firemen.

Employees in Fire Fighting

Since the danger of fire with its concomitant risk to life and property affects every employee, many experts feel that the responsibility in case of fire is a shared one. Few disagree that everyone must be educated in the principles of fire prevention and fire protection, including indoctrination in evacuation procedures and how to report a fire. But beyond this there is little agreement on what employees should be asked to do.

Some business offices set up a system of floor wardens whose job is to pass the word for evacuation and who then sweep their area of responsibility to see that it is clear of personnel, that papers are deposited in fire-proof containers and that high value, portable assets are removed from the premises.

Others take the view that their employees were not hired to act as emergency supervisors and do not expect them to act as such. Many firms of this latter persuasion ask that certain minimal functions be performed by those persons who are on hand but do not assign roles to specific people. Examples of this might be a policy of returning tapes to a fire-proof container in the computer area or securing fire-resistant safes in the accounting or cashier's office before evacuation. This responsibility would fall on the personnel in these areas at the time of the alarm and would or should take little time to accomplish. When the signal for evacuation is given, no time should be lost in vacating the building.

Many professionals feel that office employees should never be asked to do more than see to their own safety by beating an orderly retreat along predetermined escape routes. Only in the most extreme emergency—and then only if they are otherwise trapped—should employees engage in fighting a fire of any magnitude. They can be expected as a normal reaction to make an effort to put out a wastebasket fire or a small blaze in a broom closet, but even in these cases the alarm must be given as first priority. Any fire that threatens to involve a major part of an office or other parts of the building should be left to professionals.

Obviously, all such situations are matters of on-the-spot judgment. Policies concerning every situation are difficult, if not impossible, to predetermine.

Industrial Fire Brigades

The situation is quite different in industrial fire operations. In such facilities, the formation of a fire brigade composed of a few selected and trained employees is fairly general practice. There is general agreement that the nature of their employment in industrial areas makes these employees more competent to handle fire-fighting assignments, which are in many cases not that far removed from their regular work.

The exact size of each fire protection organization will vary according to the size and nature of each plant. Very large facilities, or those whose fire risk is high because of the nature of the operation, may have a full-time fire department. In smaller or less hazardous facilities, regular employees are organized into a fire brigade, which is broken down usually by departments or areas into fire companies. These companies are assigned to a given area for purposes of fire protection, fire prevention, and fire fighting. They are also available as a part of the fire brigade in any other area of the plant if a fire occurs that requires more manpower than the assigned company can handle.

The size of the brigade will depend upon the size of the plant and the nature of the risk involved. It will also be affected by the general availability, size, competence, and response capability of the public fire-fighting facilities in the neighboring areas. Whatever the size, however, it must consist of men sufficiently well-trained and familiar with the plant operation and layout to fight fires effectively in any part of the facility if the need arises.

The plant engineer and the maintenance crew should certainly be included in the brigade. Their knowledge in servicing valves, pumps, and other machinery is invaluable in an emergency situation.

The Brigade Chief

The brigade must have a clear chain of command, headed by a chief who is qualified by experience, training, or background in an allied field, or preferably all three. He will command the brigade in all fire operations.

The brigade chief will establish training programs for brigade members; he will maintain contact with local fire departments; and he will have clear authority in all matters where fire risk is concerned, including the storage of particularly flammable materials, fire fighting equipment, fire escapes, and fire doors. He will be involved in the planning of new construction or remodeling of existing facilities. He will appoint an assistant chief to act in his absence, and captains for each of the fire companies in the brigade. He will additionally establish procedures for the inspection and maintenance of all fire fighting equipment on a regular basis.

Finally, his assignment as brigade chief should be his primary duty. Members of the brigade, unless it is a proprietary fire department exclusively, will be employees who are assigned to fire protection as a collateral duty; the chief should have such duty as his primary responsibility.

Evacuation

Evacuating Industrial Facilities

Evacuation plans for an industrial facility are relatively simple, since most buildings within the perimeter are one-, two-, or in rare cases three-story buildings. Because they are occupied by personnel of a single company or under that company's jurisdiction, a single plan involving all personnel can be drawn up. And because, in most cases, aesthetic considerations are not a prime concern in the design of the industrial building, fire escapes can be constructed in any way for the greatest safety.

Although many of these buildings have elevators, most of which serve the dual purpose of hauling freight and personnel, these elevators are not necessarily the prime means of moving to and from the upper level, as they are in a high-rise office building.

Generally these buildings can be cleared in minutes. This is not to say that an evacuation plan is not needed. It is—and it must be widely distributed and clearly understood—but because most industrial buildings are more open, the exits are more visible—more a part of the unconscious orientation of the employees and, therefore, more a part of the natural traffic flow than in many other types of construction.

Evacuating High-Rise Buildings

Evacuation from high-rise urban buildings is quite a different story. In this situation, employees come to work and leave regularly by elevator. Yet if there should be a fire, they are told they must *not* use the only means of entrance or exit they really know.

In most cases they have never been on the fire stairs. They may have only a vague idea where those stairs are located, but in a time of emergency—a time of anxiety bordering on panic, when instinctual behavior would be most natural—they are asked to vacate the premises in a way which to them is very unnatural indeed.

Even in wide-open industrial facilities where orientation is quick and easy, there will always be some people who will pass a clearly marked fire exit to get to the employee door they are used to using. How much more likely this is to happen in buildings with windowless corridors and fire exits, however clearly lighted and marked, well off the normal traffic pattern used by the employees!

To overcome this problem—which must be overcome until such time as elevators are made safe to use in the case of fire—it is advisable, as a drill, to walk every employee from his desk to the nearest fire exit and to the nearest secondary exit which he would use in case his first escape route was cut off by the fire or smoke.

This could be done over a period of time with small groups at each drill. It is important that the drill actually start at the desk or office of the employee, so that the route as well as location of the exit is made clear.

Planning and Training

Evacuation plans must be based on a well-considered system and on thorough and continuing education. They should also be based on indoctrinating employees in the principles of fire safety and stressing that they are to make their own way to the proper exit and leave as quickly and calmly as possible.

Adults do not respond to being lined up like children at a fire drill and marched down the fire stairs. Whereas they might be inclined to follow a leader under many circumstances, when it comes to a concept as simple as vacating the premises, a leader has no purpose or place. They will rebel or even panic if they feel restrained or regimented in their movements toward the exit.

In setting up plans for evacuation, it might be well to review and evaluate the circumstances of a given facility and then ask a few questions:

· Are routes to exits well lighted, fairly direct, and free of obstacles?

- Are elevators posted to warn against their use in case of fire? Do these signs point out the direction of fire exits?

- Are handicapped persons provided for?

- Do corridors have emergency lighting in the event of power failure?

- Who makes the decision to evacuate? How will personnel be notified?

- Who will operate the communication system? What provisions have been made in case the primary communication system breaks down? Who is assigned to provide and receive information on the state of the emergency and the progress of the evacuation?   By what means?

PLANNING SECURITY

Security by Design

No business exists without a security problem of some kind, and no building housing a business is without security risk. Yet few such buildings are ever designed with any thought given to the steps that must eventually be taken to protect them from criminal assault.

A building must be many things in order for it to satisfy its occupant. It must be functional and efficient; it must achieve certain aesthetic standards; it must be properly located and accessible to the markets served by the occupant; and it must provide security from interference, interruption, and attack. Most of these elements are provided by the architect—but all too frequently the important element of security is overlooked.

Good security requires thought and planning—a carefully integrated system. Most security problems arise simply because no one has thought about them.

This is especially true where a company building is concerned. Since few architects have any training or knowledge in security matters, they design buildings that assist the burglar or the vandal by doing nothing to deter him. The clients themselves seldom consider security in the planning stage, and buildings are erected which provide needless opportunities for crime.

Once the building has been constructed, the damage has been done. Security weaknesses begin to manifest themselves, but it is far too expensive to make basic structural changes to correct them. Guard services and protective devices that might not otherwise have been necessary must be instituted. In any event, there will be some considerable expense for protection that could easily have been incorporated into the design of the building before construction began. This kind of oversight can be very expensive indeed.

Unfortunately, we have not yet arrived at the point where the need for security from criminal acts is as automatic a consideration as the need for efficiency or of profits. The architect's interest in design for protection of buildings and grounds is usually minimal. He traditionally leaves such demands to the client, who usually is unaware of the availability of protective hardware and is rarely competent to deal in the problems of protective design.

The rising crime rate and the growing awareness of the problem has, however, directed more attention toward the important role that building design can play in security. There now appear to be some efforts on the part of the federal government to accentuate the architect's role in security.

Security Principles in Design

There are certain principles which should always be considered in planning any building; without them it can be dangerously vulnerable.

- The number of perimeter and building openings should be kept to a minimum consistent with safety codes.

- Perimeter protection should be planned as part of the overall design.

- Exterior windows, if they are less than 14 feet above ground level, should be constructed of glass brick, laminated glass or plastic materials, or shielded with heavy screening or steel grills.

- Points of possible access or escape which breach the exterior of the building or the perimeter protection should be protected. Points to be considered are skylights, air conditioning vents, sewer ducts, manholes, or any opening larger than 96 square inches.

- High quality locks with readily changeable cylinders should be employed on all exterior and restricted area doors for protection and quick-change capability in the event of key loss.
- Protective lighting should be installed.
- Shipping and receiving areas should be widely separated.
- Exterior doors intended for emergency use only should be alarmed.
- Exterior service doors should lead directly into the service area so that non-employee traffic is restricted in its movement.
- Dock areas should be designed so drivers can report to shipping or receiving clerks without moving through storage areas.
- Employment offices should be located so that applicants either enter directly from outside or move through as little of the building as possible.
- Employee entrances should be located directly off the gate to the parking lot.
- Employee locker rooms should be located by employee entrance and exit doors.
- Doors in remote areas should be alarmed.

Chapter 3

INTERNAL THEFT

It is sad but true that virtually every company will suffer losses from internal theft—and these losses can be enormous. "A well-informed security superintendent of a nationwide chain of retail stores recently estimated that it takes between forty and fifty shoplifting incidents to equal the annual loss caused by one dishonest individual inside the organization."[5] Some estimates fix the annual loss due to internal theft at $6 billion nationwide. Obviously a problem of such magnitude must be vigorously dealt with.

THE DISHONEST EMPLOYEE

Unfortunately there is no sure way by which potentially dishonest employees can be recognized. Proper screening procedures can eliminate applicants with an unsavory past or those who seem unstable and therefore possibly untrustworthy. There are even tests that purport to measure an applicant's honesty index. But tests and employee screening can only indicate potential difficulties. They can screen out the more obvious risks, but they can never truly vouch for the performance of any prospective employee under the circumstances of new employment or under changes that may come about in his life apart from his job.

Why Employees Steal

Since there is no fail-safe technique for recognizing the potentially dishonest employee on sight, it is important to try to gain some insight into the reasons they may steal. If some rule of thumb can be developed which will help to identify the patterns of the potential thief, it would provide some warning for an alert manager.

There is no simple answer to the question of why heretofore honest men and women suddenly start to steal from their employer.

The mental and emotional process that leads to this is complex, and motivation may come from any number of sources.

The Opportunity to Steal

A high percentage of employee thefts begin with the opportunities that are regularly presented to them. If systems are lax or supervision is indifferent, the temptation to steal that which is improperly secured or unaccountable may be too much to resist by any but the most resolute employees.

Many experts agree that the fear of instant discovery is the most important force in deterring internal theft. When that likelihood is eliminated, theft is bound to follow. Threats of dismissal or prosecution of any employee found stealing are never as effective as the conviction that management supervision is such that discovery will almost certainly follow any theft.

Reason for Stealing

Some employees steal because of resentment of real or imagined injustice which they blame upon management indifference or malevolence. Some feel that they must maintain status and steal to augment their income after financial problems. Some may steal simply to tide themselves over in a genuine emergency; they rationalize the theft by assuring themselves that they will return the money after the current problem is solved. Some simply want to indulge themselves, and many, strangely enough, steal to help others.

Danger Signs

The stated and even the more deeply seated root causes of such thefts are many and varied, but there are certain signs that can indicate that a hazard exists.

The conspicuous consumer is perhaps the most obvious and the easiest risk to identify. An employee who habitually or suddenly is seen in flashy cars, flashy clothes and with flashy companions is a man to watch. His habits are visibly extravagant and he gives the impression of having found a money tree growing in his back yard. Even though he may not be stealing to support his expensive tastes, he is likely to run into financial difficulties at his pace; then he may feel obliged to look beyond his salary check to support his life style.

Any employee who shows a pattern of financial irresponsibility is a potential risk. Many people are simply incapable of handling their own affairs. They may do their job with great skill and efficiency, but they are in constant difficulty in their private lives. These people are not necessarily compulsive spenders, nor do they necessarily have expensive tastes. (They probably live quite modestly, since they have never been able to manage their affairs effectively enough to live otherwise.) They are simply people unable to come to grips with their own economic realities.

Garnishments or inquiries by creditors may identify such an employee for you. If there seems a reason to make one, a credit check might reveal the tangled state of affairs.

The employee caught in a genuine financial squeeze is also a possible problem. If he has been hit with financial demands from an illness in the family or possibly a heavy tax lien, he may find the pressures too great to bear. If such a situation comes to the attention of management, counseling is in order. Many companies maintain funds which are designed to make low interest loans in such cases; alternatively, some arrangement might be worked out through a credit union. In any event, an employee in such extremities needs help fast. He should get it both as a humane response to his needs and as a means of protecting company assets.

In addition to these general categories, there are specific danger signals which should be noted:

- Gambling on or off premises.
- Excessive drinking or signs of alcoholism.
- Obvious extravagance.
- Persistent borrowing.
- Requests for advances.
- Bouncing personal checks or checks post-dated.

What Employees Steal

Theft of Cash

The employee thief will take anything that he may consider useful to him or that has a resale value. He can get at the company funds in many ways—directly or indirectly—through collusion with vendors, collusion with outside thieves or hijackers, fake invoices, receipting for goods never received, falsifying inventories, payroll padding, false certification of overtime, padded expense accounts, cash register manipulation, overcharging, undercharging or simply by gaining access to a cashbox.

This is only a sample of the kinds of attack that are made on company assets using the systems set up for the operation of the business. It is in these areas that the greatest losses can occur, since they are frequently based on a systematic looting of the goods and services in which the company deals, and the attendant operational cash flow.

Theft of Company Property

Significant losses do occur, however, in other, sometimes unexpected areas. Furnishings frequently disappear. In some firms with indifferent traffic control procedures, this kind of theft can be a very real problem. Desks, chairs, paintings, rugs—all can be carried away by the enterprising employee thief.

Office supplies can be another problem if they are not properly supervised. Beyond the anticipated attrition in pencils, paper clips, note pads and rubber bands, these materials are often stolen in case lots. Many firms which buy their supplies at discount are, in fact, receiving stolen property. The market in stolen office supplies is a brisk one and becoming more so as the prices for this merchandise soar.

The office equipment market is another active one, and the inside thief is quick to respond to its needs. Typewriters always bring a good price, and with miniaturization, calculators and mini-computer units make tempting and easy targets.

Theft of Personal Property

Personal property is also vulnerable. Office thieves do not make fine distinctions between company property and that of their fellow workers. The company has a very real stake in this kind of theft, since personal tragedy and decline in morale follow in its wake.

Although security personnel cannot assume responsibility for losses of this nature, since they are not in a position to know about the property involved or to control its handling (and should so inform all employees), they should make every effort to apprise all employees of the threat. They should further note from time to time the degree of carelessness the staff displays in its handling of personal property and send out reminders of the potential dangers of loss.

Methods of Theft

Since it is estimated that somewhere between 7 and 10 percent of business failures annually are the result of some form of employee dishonesty, there is a very real need to examine the shapes it frequently takes. There is no way to describe every kind of theft, but some examples here may serve to give some idea of the dimensions of the problem.

- Payroll and personnel employees collaborating to falsify records by the use of nonexistent employees or by retaining terminated employees on the payroll.
- Padding overtime reports, part of which extra unearned pay is kicked back to the authorizing supervisor.
- Pocketing unclaimed wages.
- Splitting increased payroll which has been raised on checks signed in blank for use in authorized signer's absence.
- Maintenance personnel and contract servicemen in collusion to steal and sell office equipment.
- Receiving clerks and truck drivers in collusion on falsification of merchandise count. Extra unaccounted merchandise is fenced.
- Purchasing agents in collusion with vendors to falsify purchase and payment documents. Purchasing agent issues authorization for payment on goods never shipped after forging receipt of shipment.
- Purchasing agent in collusion with vendor to pay inflated price.
- Mailroom and supply personnel packing and mailing merchandise to themselves for resale.
- Accounts payable personnel paying fictitious bills to an account set up for their own use.
- Taking incoming cash without crediting the customer's account.
- Paying creditors twice and pocketing the second check.
- Appropriating checks made out to cash.

- Raising the amount of checks after voucher approval or raising the amount of vouchers after their approval.

- Pocketing small amounts from incoming payments and applying later payments on other accounts to cover shortages.

- Removal of equipment or merchandise with trash.

- Invoicing goods below regular price and getting a kick-back from the purchaser.

- Underringing on a cash register.

- Issuing (and cashing) checks on returned merchandise not actually returned.

- Forging checks, destroying them when returned with statement from the bank, and changing cash books accordingly.

These are a few of the techniques that have been used to defraud an employer. Each one is workable. Unless a concerted effort is made to control them, they or others like them can destroy a company.

Controlling Employee Theft

When we consider that internal theft accounts for five times the loss caused by automobile thieves, burglars and armed robbers combined, we must be impressed with the scope of the problem facing today's businessman. Fortunately, there are steps that can be taken to control internal theft. Losses can be cut to relatively insignificant amounts by setting up a program of education and control which is vigorously administered and supervised.

Management Failure

Many security specialists have speculated that the enormous losses in internal theft come largely from firms which have refused to believe that their employees would steal from them. As a result, their loss prevention systems are weak and ineffective.

Managers of such companies are not naive. They are aware that the criminally inclined employee exists—and in great numbers—but he exists in other companies. Such men are truly shocked when they discover how extensive employee theft can be. This doesn't mean that a high percentage of the employees are involved (though they may be). Two or three employees, given the right opportunities, can create havoc in any business in a very short time—and the great majority of them go undetected.

National crime figures do not necessarily concern each individual manager. He can shrug them off as an enormous problem that someone else must deal with, one that has no effect on his operation—and hopefully he is right. But he simply cannot afford to close his eyes to the potential damage that he faces. He must be convinced that one of his employees *might* steal from him. He must be persuaded that there is, at least, a *possibility* that he is exposed to embezzlement and that maybe sooner or later someone will be unable to resist the temptation to take what is so alluringly available.

It may be surprising to find that there are shrewd businessmen who are still willing to ignore the threat of internal theft and who are therefore unwilling to take defensive steps against it. But the fact is they do exist in large numbers. Many security men report that the biggest problem is in convincing management that the problem is there and that steps must be taken for the protection of employer and employee alike.

There is some evidence and more speculation that these managers are reluctant to, in effect, declare themselves as suspicious of the employees—to make overt action implying distrust of men and women, many of whom are perhaps old and trusted members of the corporate family. Such an attitude does the manager honor but it is somewhat distorted. It fails to take into account the dynamics of all our lives. The man who today would stand firm against the most persuasive temptation, tomorrow finds himself in different circumstances—perhaps with different resultant attitudes. Whereas today the thought of embezzlement is repugnant to him, tomorrow he might treat the notion with a different view.

The Contagion of Theft

Theft of any kind is a contagious disorder. Petty, relatively innocent pilferage by a few spreads through the facility. As more people participate, more will participate, until even the most rigid break down and join in. Pilferage becomes acceptable—even respectable. It has a general social acceptance which is reinforced by almost total peer participation. Few men make independent ethical judgments under such circumstances. In this microcosm, the act of petty pilferage is no longer viewed as unacceptable conduct. It has become, not a permissible sin, but a right.

The docks of New York City were an example of this progression. Forgetting for the moment the depredations of organized crime and the climate of dishonesty that characterized that operation for so many years, even longshoremen not involved in organized theft had worked out a system all their own. For every so many cases of whisky unloaded, for example, one case went to the men. Little or no attempt was made to conceal this. It was a tradition, a right. When efforts were made to curtail the practice, labor difficulties arose. It soon became evident that certain pilferage would have to be accepted as an unwritten part of the union contract under the existing circumstances.

This is not a unique situation. The progression from limited pilferage to its acceptance as normal conduct to the status of an unwritten right has been repeated time and again. The problem is, it doesn't stop there. Ultimately pilferage becomes serious theft, and then the real trouble starts.

Even before pilferage expands into larger operations, it presents a difficult problem to any business. Even where the goods taken by any one individual is small, the aggregate can represent a significant business expense. With the costs of materials, manufacture, administration and distribution rising as they are, there is simply no room for added, avoidable expenses in today's competitive markets. The business that can operate the most efficiently, and offer quality goods at lower prices because of the efficiency of its operation, will have a huge advantage in the marketplace. When so many companies are fighting for their very economic life, there is simply no room for waste—and pilferage is just that.

Moral Obligation to Control Theft

It is also important to observe that management has a moral obligation to its employees to protect their integrity by taking every possible step to avoid presenting open opportunities for pilferage and theft that would tempt even the staunchest of men to take advantage of the trust placed in them.

This is not to suggest that each company should assume a paternal role toward its employees and undertake their responsibilities for them. It is to suggest strongly that the company should keep its house sufficiently in order to avoid inciting employees to acts that could result in great personal tragedy as well as in damage to the company.

PLANNING FOR INTERNAL SECURITY

A plan for internal security must have the whole-hearted support and involvement of management from the highest echelon. Any initial reluctance to institute such a plan must be overcome with tact and conviction. Once that has been accomplished, the nature of the commitment must be fully understood. No security program, or any other, for that matter, can be effective without management's full and continuing support.

This aspect of the security program, like every other one, must be integrated into the whole which, as frequently noted, must be a total system involving all aspects of the facility's security needs. Neither internal security nor access control nor any other part of the security program can be dealt with as an independent unit. Each must be a part of a cohesive whole in the development of a systems approach to the objective.

Personnel

The objective of a program encompassing internal security is to prevent theft by employees. If all the employees were of such character that they could not bring themselves to steal, the security personnel would have little to do. If, on the other hand, thieves predominate in the mix of employees, the system will be sorely tried, if indeed it can be effective at all. Basic to its effectiveness is the cooperation of that majority of honest personnel who perform as assigned and in so performing refuse to initiate or collaborate in conspiracies to steal. Without this dominant group, the system is in trouble from the start.

Personnel Screening

The best place to start, then, is in the personnel office, where bad risks can be screened out on the basis of reasonable security procedures.

In some industries, especially those with high technical requirements, this can be a problem, since qualified personnel may be difficult to find. There can be resistance from the employment office to the disqualification of an otherwise qualified applicant on marginal grounds of security. Here a job of education must be undertaken to convince those objectors that a man who may later embezzle from the company is a poor risk from many viewpoints, no matter how highly qualified he may be in the specific skills for which he has been considered.

Rejection of bad risks must be on the basis of standards that have been carefully established in cooperation with the personnel director. Once established, these standards must be met in every particular, just as proficiency standards must be met. Obviously the standards must be reviewed from time to time to avoid dealing with applicants unjustly and to avoid placing the company at a competitive disadvantage in the labor market by demanding more than is available. Even here, however, a bottom line must be drawn. After a certain point, compromises and concessions can no longer be made without inviting damage to the company.

Such a careful, selective program may add some expense to the employment procedure, but it can pay for itself in reduced losses, better people and lower turnover. And the savings in crimes that never happened, though unknowable, could be thought of as enormous.

Employment History

It is important that the application form ask for a chronological listing of all previous employers, among other things, in order to provide a list of firms to be contacted for information on the applicant as well as to show continuity of career. Any gaps could indicate a jail term which had been overlooked in filling out the application. When checking with previous employers, dates on which employment started and terminated should be verified.

References submitted by the applicant must be contacted, but they are apt to be biased—after all, they were names submitted by the man to be investigated; they are not likely to be negative or hostile to his interests. It is important to contact someone—preferably an immediate supervisor—at each previous employer, and such contact should be made by phone or in person.

The usual and easiest system of contact is by letter, but this leaves much to be desired. The relative impersonality of a letter, especially one in which a form or evaluation is to be filled out, leads to impersonal and essentially uncommunicative answers. Since many companies as a matter of policy, stated or implied, are reluctant to give a man a bad reference except in the most extreme circumstances, a letter reply to a letter will, in some cases, even be misleading.

On the other hand, phone or personal contacts may become considerably more discursive and provide shadings in the tone of voice that can be important and, even when no further information is forthcoming, indicate that a more exhaustive investigation is required.

Backgrounding

It may be desirable to get a more complete history of a prospective employee—especially in cases where sensitive financial or supervisory positions are under consideration. Professional services providing this kind of investigation involve extra expense, but in many cases it can be well worthwhile. This backgrounding, which involves a discreet investigation into the past and present activities of the applicant, can be most informative.

Since it is estimated that as much as 93 percent of all persons *known* to have stolen from their employers are not prosecuted, a thorough investigation is certainly justified into the background of anyone considered for a job in which he will be responsible for significant amounts of cash or goods, or will be in a management position responsible for shipping, receiving, purchasing, or paying.

It is also significant to note that there is agreement among personnel and security experts that 20 percent of a given work force is responsible for 80 percent of the personnel problems of all kinds. If backgrounding can turn up this kind of record, it is well worth the expense.

Some firms state on their application form that applicants will be bonded or fingerprinted or must give permission to undergo a polygraph test. Any criminal with a record will probably bow out at this point, since he knows his past is likely to disqualify him.

Backgrounding is also employed to investigate employees being considered for promotion to positions of considerable sensitivity and responsibility. Such a man may have been the very model of rectitude at the time of his employment but may since have fallen upon financial reverses that threaten his life style. If a background investigation uncovers such information, a company is in a position to offer assistance if it so desires, thus relieving him of the strain and need, both of which can lead to embezzlement. Such action can boost company morale as well as reduce the potential for theft out of desperation.

Polygraph and PSE

Another approach to full background investigation is the use of the polygraph or of the psychological stress evaluator (PSE). In states where their use is legal, these machines can be useful tools in determining the past and current record of applicants for employment or promotion to positions of considerable sensitivity.

The use of this technique is controversial, as is the discussion among practitioners of the relative merits of each instrument. However, many security people look on them as invaluable tools and generally agree that their use in the hands of a competent, trained professional can be most constructive.

These instruments differ. The polygraph measures the response of the heartbeat or pulse, the skin's galvanic action, and respiration. Sensors picking up these responses are attached painlessly to the subject by wires that run directly between him and the machine. He is questioned by the use of control questions, irrelevant and relevant inquiries. The readout of his pulse, respiration and skin action is recorded on a paper tape in the machine, and the responses to various questions are analyzed in terms of his charted reactions as compared with his reactions to simple test questions which establish his normal response to lying.

The PSE works in a similar way operationally, but in this case a recording is made of the subject's replies to the questions put to him. These recordings analyze and reproduce the sound wave characteristics of the voice. By running these tapes through the instrument at different speeds, different aspects of the sound wave activity can be recorded on a graph for evaluation. Certain patterns produced by the autonomic (involuntary) nervous system indicate the truth or falsity of a given response.

Adherents of both systems claim a high degree of accuracy in such tests when they are conducted under the supervision of properly trained personnel.

It is important to determine the limitations on the use of such instruments in the various states. Some states forbid their use as a *requirement* for employment but permit them to be used on a voluntary basis. Other states—Massachusetts, for example—forbid their use in any circumstance. There is a growing controversy over the use of such machines, and many states that exercised indifferent or no supervision over their employment are enacting legislation in which certain minimum training and qualification requirements are established.

Generally speaking, organized labor has lobbied diligently for legislation banning the use of so-called lie detectors in all industrial or commercial applications, and their efforts have borne some fruit in many states. The American Polygraph Association, while opposed to the use of the PSE essentially on the grounds that it has not yet proven itself, has endorsed much legislation setting stricter standards for polygraph operators but has, of course, fought labor's stand on the matter.

Many firms do, however, make use of polygraph or PSE examinations in hiring some of their personnel, and they make wide use of the approach in investigations of various kinds. Those firms that do use these machines have, in general, found that they have served a useful purpose.

Since such examinations are expensive, costing from $50 to $150, depending on the length of the test, not every firm will find them practical. Even those firms that do find the expense acceptable usually limit their application to persons being hired or promoted to areas of particular sensitivity and responsibility. Each company must individually determine the need for such examinations, and their practicality, after they have decided whether the risk is such that the expense is warranted, and whether they can be satisfied by an evaluation based on other, cheaper methods.

Whatever the decision, the security manager would be well advised to consult a reputable firm handling polygraph examinations as well as a PSE operation to learn more about what such examinations involve.

Continuity of Program

When a company has made a systematic and conscientious effort to screen out dishonest, troublesome, incompetent and unstable employees, it has taken a first and significant step toward reducing internal theft. It is important, then, that the program continue in effect on a permanent basis.

Care must be taken to avoid relaxing standards or becoming more superficial in checking applicants. There is a tendency to lose sight of the full dimensions of the problem if the security program makes substantial inroads on the loss factor. The past is too soon forgotten, and carelessness follows close behind. Active supervision is always necessary to maintain the integrity of this important aspect of every security program.

Hiring Ex-Convicts and Parolees

It should be strongly noted here that rigid exclusionary standards should never be applied to ex-convicts or parolees who openly acknowledge their past records. Such a policy would be at best unjust and at worst irresponsible. These men have done their time. Their records are available in situations where employment is being sought. To turn them away simply on the basis of their past mistakes would be to force them into a criminal pattern of life to survive.

Some such men might well be unacceptable in certain companies—but a rigid policy denying all of them employment would be to deny the company many potentially good men who deserve a chance for rehabilitation. Experience has shown that these men, knowingly hired in the right positions and properly supervised, are not only acceptable but frequently highly responsible and trustworthy. They should be given an opportunity to re-establish themselves in society.

Morale

In any organization which exists by the cooperative efforts of all its members, it is important that each one of those members feel that he is an important part of the operation, both as a contributor and as an individual.

This dual role is important to recognize. Each employee must feel that he is a significant, integrated part of the whole and identified with it, while still maintaining and protecting his basic importance as an individual apart from the structure. If he is denied reinforcement in either of these views, his sense of personal worth suffers; anger, anxiety, frustration or feelings of inadequacy may follow. Any of these attitudes are damaging both to him and to the organization as a whole, and they must be headed off.

The best way to do that is to recognize that each employee is in fact an important member of the organization, else he wouldn't be there. A function must be performed as a part of the larger function, and he is the one performing it. He is unique in that he is unduplicatable, and he is important in that uniqueness in that he is fellow man. His reinforcement comes from management, supervisors and peers. He is a vital, irreplaceable organ of the greater body. Of such recognition and respect is high morale created and maintained, and everyone benefits.

Every supervisor must be indoctrinated in the importance of the morale factor in every business. Just as no business can operate without employees, none can operate efficiently with a work force whose morale has been damaged and who respond with an attitude of listlessness and disinterest. Threats and coercion will not restore the optimum level of morale any more than directives and public notices can. This is a condition that comes from a man's very being; it must be dealt with in those terms.

Additional factors are also important in this regard. Physical surroundings and appropriate rewards are part of the message that tells a man what the organization thinks of him.

In an atmosphere of concern, fairness and mutual respect, few men would be led to steal. It would be like stealing from themselves. It is important that such an atmosphere be developed and maintained for the good of all.

A program aimed at improving or maintaining employee morale might contain some or all of the following elements:

- Clear statements of company policy which is consistently and fairly administered.
- Regular review of wages and wage policy, updated to assure equitable wage levels.
- House organ or newsletter and bulletin boards kept current.
- Open, two-way avenues of communication between management and all employees.
- Clear procedures, formal and informal, for airing grievances and personal problems with supervisors.
- Vigorous training programs to improve job skills and pave the way for advancement.
- Physical surroundings—decor, cleanliness, sound and temperature control, and general housekeeping at a high level.

Survey Program

As in other surveys, the first requirement before setting up protective systems for internal security is to survey every area in the company to determine the extent and nature of the risks. If such a survey is conducted energetically and exhaustively, and its recommendations for action are acted upon intelligently, significant losses from internal theft will be a matter of history.

Who Conducts the Survey

Such a survey should normally be undertaken by security personnel. In instances where losses have been particularly severe and troublesome, it might be wise to call in an independent security firm to work with staff security people. A separate survey of accounting procedures, including the flow of all documents authorizing and receipting shipments, inventorying, etc., might be assigned to a specialist in operational audits, probably one who works for the same firm that handles the company's annual audit. It must be remembered, however, that this survey is to examine the operation of accounting functions and should not be considered as in any way related to normal auditing procedures.

Whoever conducts the survey must be fully familiar with all aspects of the operation in as much detail as possible. The pressures and tensions of a business should be examined as part of such a survey, as should the structure of both the formal and informal organization. The latter can be most easily identified as that unofficial chain of command or flow of work, or that acknowledged but unofficial series of power centers that operate sometimes parallel or sometimes even in opposition to the official organization of the company. It is easy to spot potential points of collusion when examining the organization chart; it is frequently less easy to spot such points in the informal organization, unless that aspect of the business is well known to the surveyor and thoroughly understood by him.

What is Covered

The survey should be conducted in a methodical manner, moving from department to department and covering every area of operation in each department before moving on. The basic question underlying this inspection is the loss potential inherent in each aspect of the routine. The greater the exposure, the more exacting the security procedures must be.

In the early stages, the survey team must try to think the way a thief would think. It must probe the routines currently in effect to find the weak links—to find those spots where a thief could strike with the greatest assurance of success and the least risk of exposure. Each member of the team must ask himself, "If I were so inclined, how would I go about stealing at this point or in that job?" This questioning and probing must also encompass every *possible* loss to theft and must not limit itself to considerations of likely areas of loss. Anything less than a total consideration of the possibilities will leave the job to that degree undone.

Since security, to be effective, requires a total systems approach, it must include every aspect of the business in its surveys. The survey for the control of internal theft is no exception. Even though its recommendations will be incorporated into the overall security plan, and even though each department will be surveyed on its own and as a part of the total picture, each aspect of the business in any way affecting its internal security must be inspected for security flaws or weaknesses that could lead to employee theft.

As a practical matter, the company need not be reinspected for each aspect of the overall security plan, but it is important to consider different categories of danger individually before synthesizing them into a single protective procedure. The survey would include an examination of physical security, for example, or of the state of fire safety, or the condition and coverage of the alarm system, but at this stage each of these elements should be approached as separate problems. There will be time enough to coordinate the needs of each when all their requirements are consolidated into a single cohesive system.

In examining the needs of internal security, it will be necessary to consider hiring and promotion practices, the handling of receipts and disbursements, credits and rebates. Petty cash, payroll, bank deposits, handling of cash and negotiables, inventories, warehousing, data processing, purchasing, supplies, receiving, shipping, merchandise controls, inspection procedures, employee badging or identification policy, time card and key control, confidential data controls and any other area or function where employee dishonesty could exist must be evaluated.

This evaluation will review the entire product of the survey, taking special note of those areas observed or surmised to be vulnerable to criminal manipulation. Such weaknesses must be eliminated or controls developed to protect against any possible breach of trust.

Need for Management Support

It is at this point especially that the strong support of top management may be needed most. In order to implement needed security controls, certain operational procedures may necessarily have to be changed. This will require cooperation at every level, and cooperation is sometimes hard to get in situations where a department manager feels his authority has been diminished in areas within his sphere of responsibility.

The problem is compounded when those changes determined necessary cut across departmental lines and even serve to some degree to alter intra-departmental relationships. Effecting systems under such circumstances will require the greatest tact, salesmanship and executive ability; and failing that (as a last resort), falling back on the ultimate authority vested in the security operation by top management. Any hesitation or equivocation on the part of either management or security at this point could damage the program before it has been initiated.

This does not, of course, mean that management must give security *carte blanche;* reasonable and legitimate disagreements will inevitably arise. It does mean that proposed security programs based on broadly stated policy must be given the highest possible priority. In those cases where conflict of procedures exists, some compromise may be necessary, but the integrity of the security program as a whole must be preserved intact.

Communicating the Program

The next step is to communicate necessary details of the program to all employees. Many aspects of the system may be confidential or on a need-to-know basis, but since part of it will involve procedures engaged in by most or all of company personnel, they will need to know those details in order to comply. This can be handled in a series of meetings explaining the need for security, the damaging effects of internal theft to jobs, benefits, profit sharing and the future of the company. Such meetings can additionally serve to notify all employees that management is taking action against criminal acts of all kinds, at every level, and dishonesty will not be tolerated.

Such a forceful statement of position in this matter can be very beneficial. Most employees are honest men and women who disapprove of those who are criminally inclined. They are apprehensive and uncomfortable in a criminal environment, especially if it is widespread. The longer such conduct is condoned by the company, the more they lose respect for it, and a vicious cycle begins. As they lose respect, they lose a sense of purpose. Their work suffers, their morale declines, and at best their effectiveness is seriously diminished. At worst they reluctantly join the thieves. A clear, uncompromising policy of theft prevention is usually welcomed with visible relief.

Continuing Supervision

Once a system is installed, it must be supervised if it is to become, and remain, effective. Left to their own devices, employees will soon find short cuts, and security controls will be abandoned in the process. Old employees must be reminded regularly of what is expected of them, and new employees must be adequately indoctrinated in the system they will be expected to follow.

This must be a continuing program of education if expected results are to be achieved. With a national average turnover of 50 percent of the white-collar work force, it can be expected that the office force, which handles key paper work, will be replaced at a fairly consistent rate. This means that the company will have a regular influx of new people who must be trained in the procedures to be followed and the reasons for them.

Program Changes

In some situations reasonable controls will create duplication of effort, cross-checking and additional paper work. Since each time such additional effort is required there is an added expense, procedural innovations requiring it must be avoided wherever possible. But most control systems aim for increased efficiency. Often this is the key to their effectiveness.

Many operational procedures, for a variety of reasons, fall into ponderous routines involving too many people and excessive paper shuffling. This may serve to increase the possibility of fraud, forgery or falsification of documents. When the same operational result can be achieved by streamlining the system, incorporating adequate security control, it should be done immediately.

Virtually every system can be improved, and every system should be evaluated constantly with an eye for such improvement, but these changes should never be undertaken arbitrarily. Procedures must be changed only after such changes have been considered in the light of their operational and security impact, and such considerations should further be undertaken in the light of their effect on the total system.

No changes should be permitted by unilateral employee action. Management should make random spot checks to determine if the system is being followed exactly. Internal auditors and/or security personnel should make regular checks on the control systems.

Violations

Violations should be dealt with immediately. Any management indifference to security procedures is a signal that they are not important, and, where work-saving methods can be found to circumvent such procedures, they will be. As soon as any procedural untidiness appears and is allowed to continue, the deterioration of the system begins.

It is well to note, too, that, while efforts to circumvent the system are frequently the result of the ignorance or laziness of the offender, a significant number of such instances are the result of an employee probing for ways to subvert the controls in order to divert company assets to his own use.

SPECIFIC CONTROLS
Auditing Assets

Periodic audits by outside auditors are essential to any well-run security program. Such an examination will discover theft only after the fact, but it will presumably discover any regular scheme of embezzlement in time to prevent serious damage. If these audits, which are normally conducted once a year, are augmented by one or more surprise audits, even the most reckless criminal would hesitate to try to set up even a short-term scheme of theft.

These audits will normally cover an examination of inventory schedules, prices, footings and extensions. They should also verify current company assets by physical inventory sampling, accounts receivable, accounts payable (including payroll), deposits, plant and outstanding liabilities. In all these cases a spot check beyond the books themselves can help to establish the existence of legitimate assets and liabilities, not empty entries created by a clever embezzler.

Cash

Any business handling relatively few cash payments in and out is fortunate indeed. Such a business is able to avoid much of the difficulty created by this security sensitive area, since cash handling is certainly the operation most vulnerable and the most sought after by the larcenous among the staff.

Cash by Mail

If cash is received by mail—a practice which is almost unheard of in most business—its receipt and handling must be undertaken by a responsible, bonded supervisor or supervisors. This administrator should be responsible for no other cash handling or bookkeeping functions. This official should personally see to it that all cash received is recorded by listing the amount, the payer, and such other pertinent information as procedures have indicated. This list should be made, in duplicate, on sequentially numbered forms, with both copies signed by the manager in charge of opening the mail as well as by the cashier who receipts for the money. The cashier will then keep one copy of the record for file—the other will be forwarded to accounting. Both will verify the numbering on the cash receiving list.

There is clearly a danger here at the very outset. If cash is diverted before it is entered on any receipt, there is no record of its existence. Until it is channeled into company ledgers in some way and begins its life as a company asset, there is no guarantee that it won't serve some more private interest. This requires supervision of the supervisor. In the case of a firm doing a large catalogue business, receiving large amounts of cash in spite of pleas for check or money orders, it has sometimes been felt that the operation should be conducted in a special room reserved for the purpose.

Daily Receipts

All cash book entries must be checked against cash on hand at the end of each day. Spot checks on an irregular basis should also be conducted.

Cash receipts should be deposited in the bank intact and each day's receipts balanced with the daily deposit. Petty cash as needed should be drawn by check.

All bank deposits should be accompanied by three deposit slips, one of which is receipted by the bank and returned to the cashier by the person making the deposit. The second is mailed to the office accounting department and the third is the bank's.

Each day's deposit slips should be balanced with the day's receipts.

Bank Statements

Bank statements must be received and reconciled by someone who is not authorized to deposit or withdraw funds, or to make a final accounting of receipts or disbursements. When bank statements are reconciled, cancelled checks should be checked against vouchers for alterations and for proper endorsement by the payee. Any irregularities in the endorsements should be promptly investigated. If the statement itself seems in any way out of order by way of erasure or possible alteration, the bank should be asked to submit a new statement to the reconciling official's special personal attention.

Petty Cash

A petty cash fund, set aside for that purpose only, should be established. The amount to be carried in such a fund will be based upon past experience. These funds must never be commingled with other funds of any kind and should be drawn from the bank by check only. They should never be drawn from cash receipts. No disbursements of any kind should be made from petty cash without an authorized voucher signed by the employee receiving the cash and countersigned by authorized authority. No voucher should be cashed that shows signs of erasure or alteration, and all such vouchers should be drawn up in ink or they should be typed. In cases of typographical error, new vouchers should be prepared rather than correcting the error. If there is any reason for using a voucher on which an erasure or correction has been made, the authorizing official should initial the change or place of erasure.

Receipts substantiating the voucher should accompany it and should, if possible, be stapled or otherwise attached to it.

The petty cash fund should be brought up to the specified amount from time to time as prearranged, by check to the amount of its depletion. The vouchers upon which disbursements were made should always be verified by an employee other than the one in charge of the fund. All vouchers submitted and paid should be cancelled in order to avoid re-use.

Petty cash should be balanced occasionally by management, at which time vouchers should be examined for irregularities.

Separation of Responsibility

The principle of separation of responsibility and authority in matters concerning the company's finance is of prime importance in security management. This situation must always be sought out in the survey of every department. It is not always easy to locate. Sometimes even the employee who has such power is unaware of his dual role, but the security specialist must be sensitive to its existence and provide for an immediate change in such operational procedures whenever they appear.

An employee who is in the position of both ordering and receiving merchandise, or a cashier who authorizes and disburses expenditures, are examples of this double-ended function in operation. All situations of this nature are potentially damaging and should be eliminated. Such procedures are manifestly unfair to company and employee alike. To the company because of the loss that might incur; to the employee because of the temptation and ready opportunity they present. Good business practice demands that such invitations to embezzlement be studiously avoided.

The Cashier

It is equally important that cash handling be separated from the record-keeping function. When the cashier becomes his own auditor and bookkeeper, he has a free rein with that part of the company funds. The chances are he won't steal, but he could—and he might. He might also make mistakes without some kind of double check on his arithmetic.

In some smaller companies this division of function is not always practical. In such concerns it is common for the bookkeeper to act also as cashier. If this is the case, a system of countersignatures, approvals and management audits should be set up to help divide the responsibility of handling company funds as well as accounting for them.

Promotion and Rotation

Most embezzlement is the product of a scheme operating over an extended period of time. Many embezzlers prefer to divert small sums on a systematic basis, feeling that the individual thefts will not be noticed and therefore the total loss is unlikely to come to management's attention.

These schemes are ultimately frustrated (when they are at all) either by some accident that uncovers their system or by the greed of the embezzler, who is so carried away by the success of his efforts that he steps up the ante. But while the theft is working and he is in control, it is usually difficult to detect. Frequently he is in a position to alter or manipulate records in such a way that the theft escapes the attention of both internal and outside auditor. This can be countered by upward or lateral movement of employees.

Promotion from within wherever possible is always good business practice, and lateral transfers can be effective in countering possible boredom or the danger of reducing a function to rote and thus diminishing its effectiveness.

Such movement also frustrates the embezzler. When he loses control of the books governing some aspect of the operation, he loses the opportunity to cover his theft. Discovery would inevitably follow a careful audit of books he could no longer manipulate. If regular transfers were a matter of company policy, no rational embezzler would set up a long-term plan of embezzlement unless he found a scheme that was audit-proof, and such an eventuality is highly unlikely.

To be effective as a security measure, such transfers need not involve all personnel, since every change in operating personnel brings with it changes in operation. In some cases, even subtle changes may be enough to alter the situation sufficiently to reduce the totality of control an embezzler has over his books. If such is the case, his swindle is over. He may avoid discovery of his previous looting, but he cannot continue in his grab without danger of being unmasked.

The Reluctant Vacationer

In the same sense, embezzlers dislike vacations. They are aware of the danger if someone else should handle their accounts, if only for the two or three weeks of vacation. So they make every effort to pass up the holiday.

Any manager who has a reluctant vacationer on his hands should recognize that he has a potential problem. Vacations are designed to refresh the outlook of everyone. No matter how tired they may be when they return to work, every vacationer has been refreshed emotionally and intellectually. His effectiveness in his job has probably improved and he is, generally speaking, a better man for it. The company benefits from his vacation as much as he does. No employee should be permitted to pass up his duly authorized vacation—especially one whose position gives him access to control over company assets.

Access to Records

Many papers, documents and records are confidential, or at least are available to only a limited number of people who need such papers in order to function. All other persons are deemed to be off limits. They have no need for the information. Such papers should be secured under lock and key—and, depending on their value or reconstructability, in a fire resistant container.

Forms

Company forms are often extremely valuable to the inside as well as the outside thief. They should be secured and accounted for at all times. If it is at all possible or feasible to do so, they should be sequentially numbered and recorded regularly so that any loss can be detected at a glance.

Blank checks, order forms, payment authorizations, vouchers, receipt forms and all others which authorize or verify transactions are prime targets for thieves and should therefore be accounted for.

Since there are many effective operational systems in use for the ordering, shipping or receipting of goods, as well as the means by which all manner of payments from petty cash to regular debt discharge are authorized, no one security system to protect against illegal manipulation within such systems would apply universally. It can be said, however, that since every business has some means to authorize transactions of goods or money, the means by which such authorizations are made must be considered in the security program. Security of such means must be considered as an important element in any company's defense against theft.

Generally speaking, all forms should be pre-numbered and, where possible, used in numerical order. Any voided or damaged forms should be filed and recorded and forms reported lost must be accounted for and explained. All such numbered forms of every kind should be inventoried and accounted for periodically.

Purchase Orders

In cases where purchase orders are issued in blocks to various people who have the need for such authority, such issuance must be recorded and disposition of their use should be audited regularly. In such cases it is customary for one copy of the numbered purchase order to be sent to the vendor, who will use that number in all further dealings on that particular order; another copy will be sent to accounting for purposes of payment authorization and accrual if necessary; and one copy will be retained by the issuing authority. Each block issued should be used sequentially, although, since some areas may have more purchasing activity than others, purchasing order copies as they are forwarded to accounting may not be in overall sequence.

Purchasing

Centralized Responsibility

Where purchasing is centralized in one department, controls will always be more effective. Localizing responsibility as well as authority reduces the opportunity for fraud accordingly. This is not always possible or practical, but in areas where purchasing is permitted by departments needing certain materials and supplies, there can be confusion occasioned by somewhat different purchasing procedures. Cases have been reported where different departments pay different prices for the same goods and services and thus bid up the price the company is paying. Centralization of purchasing would overcome this problem.

Purchasing should not, however, be involved in any aspect of accounts payable or the receipt of merchandise other than informationally.

Competitive Bids

Competitive bids should be sought wherever possible. This, however, raises an interesting point that must be dealt with as a matter of company policy. Seeking competitive bids is always good practice, both to get a view of the market and to provide alternatives in the ordering of goods and materials, but it does not follow that the lowest bidder is always the vendor to do business with.

Such a bidder may lack adequate experience in providing the services bid for, or he may have a reputation of supplying goods of questionable quality even though they may meet the technical standard prescribed in the order. A firm may also underbid the competition in a desperate effort to get the business, but then find it cannot deliver materials at that price, no matter what it has agreed to in its contract.

In order to function wisely and to be able to exercise good judgment in its area of expertise, purchasing must be permitted some flexibility in its selection of vendors. This means that it will not always be the low bidder who wins the contract.

Now, since competitive bidding provides some security control in reducing favoritism, collusion and kickbacks between the purchasing agent and the vendor, these controls would appear to be weakened or compromised in situations where the purchasing department is permitted to select the vendor on considerations other than cost. This can be true to some degree, but this is a situation in which business or operational needs may be in some conflict with tight security standards — and in which security should revise its position to accommodate the larger demands of efficiency and ultimate economy. After all, cheap is not necessarily economical.

Controls in this case could be applied by requiring that in all cases where the lowest bid was not accepted, a brief explanation in outline form be attached to the file along with all bids submitted. Periodic audits of such files could establish if any pattern of fraud seems likely. Investigation of the analysis or assumptions made by purchasing in assigning contracts might be indicated in some situations to check the validity of its stated reasoning in the matter.

Other Controls

Copies of orders containing the amount of merchandise purchased should not be sent to receiving clerks. These clerks should simply state the quantity actually received with no preconception of the amount accepted. Payment should be authorized only for that amount actually received.

Vendor invoices and receipts supporting such vouchers should be cancelled to avoid the possibility of their resubmission in collusion with the vendor.

Purchasing should be audited periodically and documents should be examined for any irregularities.

Payroll

It is important that the payroll be prepared by persons who will not be involved in its distribution. This is consistent with the effort to separate the various elements of a particular function into its component parts and then distribute the responsibility for those parts to two or more persons or departments.

Every effort should be made to distribute the payroll in the form of checks rather than cash, and such checks should be of a color different from those used in any other aspect of the business. They should also be drawn on an account set aside exclusively for payroll purposes. It is also important that this account be maintained in an orderly fashion and avoid using current cash receipts for payroll purposes.

Personnel Records

The payroll should be prepared from personnel records which in turn have come from personnel as each employee is hired. Such a record should contain basic data such as name, address, attached W-2 form, title, salary and any other information that the payroll department may need. The record will be countersigned by a responsible executive verifying the accuracy of the information forwarded.

This same procedure should be followed when an employee terminates his employment with the company. All such notifications should be consolidated into a master payroll list, which should be checked frequently to make sure that payroll's list corresponds to the employment records of personnel.

Unclaimed Checks

Unclaimed paychecks should be returned to the treasurer or controller after a reasonable period of time for redeposit in the payroll account. Certainly such cases should be investigated to determine why the checks were returned or why they were issued in the first place. All checks so returned should be marked to prevent any re-use and filed for reference. Since payrolls usually reflect overtime and other payments in addition to regular salary disbursements, such payments should be supported by time sheets authorized by supervisors or department heads. Time sheets of this nature should be verified periodically to prevent overtime padding and kickback.

Time cards themselves should be marked to prevent re-use.

Payroll Audits

The payroll should be audited periodically for any irregularities, especially if there has been an abnormal increase in personnel or net labor cost.

To further guard against the fraudulent introduction of names into the payroll, distribution of paychecks should periodically be undertaken by the internal auditor, the treasurer or other responsible offical. In large firms this can be done on a percentage basis, thus providing at least a spot check of the validity of the rolls.

Accounts Payable

As in the case of purchasing, accounts payable should be centralized to handle all disbursements upon adequate verification of receipt and proper authorization for payment.

These disbursements should always be by checks that are consecutively numbered and used in that order. Checks that are damaged, incorrectly drawn, or for any reason unusuable must be marked as cancelled and filed for audit. All checks issued for payment should be accompanied by appropriate supporting data, including payment authorizations, before they are signed by the signing authority. It is advisable to draw the checks on a check writing machine which uses permanent ink and is as identifiable to an expert as handwriting or a particular typewriter. Checks themselves should be of safety paper which will show almost any attempted alteration.

Here, as in order departments, periodic audits must be conducted to examine the records for any sign of nonexistent vendors, irregularities in receipts or payment authorizations, forgeries, frauds or unbusinesslike procedures that could lead to embezzlement.

General Merchandise

Merchandise is always subject to embezzlement, particularly when it is in a transfer stage, as when it is being shipped or received. The dangers of loss at these stages are increased in operations where controls over inventory are lax or improperly supervised.

Separation of Functions

To control these sensitive aspects of any operation involving the handling of merchandise, it is desirable to separate the three functions. Receiving, warehousing and shipping should be the responsibility of three different areas. Movement of merchandise from one mode to another should be accompanied by appropriate documents which clearly establish the responsibility for specific amounts of merchandise passing from one sphere of authority to another.

Receipting for a shipment places responsibility for a correct count and the security of the shipment on the receiving dock. This responsibility remains his until he transfers it and receives a proper receipt from the warehouse man. The warehouse man must verify and store the shipment, which is his responsibility until it is called for by the sales department, for example, or perhaps directed to be shipped by authorized voucher. The warehouse man assembles the goods, passes them along as ordered, and receives a receipt for those goods delivered.

In this process, responsibility is fixed from point to point. Various departments or functions take on and are relieved of responsibility by voucher and receipt. In this way a perpetual inventory is maintained as well as a record of responsibility for the merchandise.

All vouchers and requisitions must be numbered to avoid the destruction of records or the introduction of unauthorized transfers into the system. Additionally, stock numbers of merchandise should accompany all of its movement to describe the goods and thus aid in maintaining perpetual inventory records.

In small firms where this separation of duties is impractical and receiving, shipping and warehousing are combined in one person, the perpetual inventory is essential for security, but it must be maintained by someone other than the person actually handling the merchandise. The shipper-receiver-warehouser should not have access to these inventory records at any time.

Inventories

Inventories will always be an important aspect of merchandise control, no matter what operations are in effect. Such inventories must be conducted by someone other than the person in charge of that particular stock. In the case of department stores, for purposes of inventory, personnel should be moved to a department other than their regular assigned one.

In firms where a perpetual inventory record is kept, physical counts on a selective basis can be undertaken monthly or even weekly. In this procedure a limited number of certain items randomly selected can be counted and the count compared with current inventory record cards. Any discrepancies can be traced back to the point of loss to determine its cause.

Physical Security

It is important to remember that personnel charged with the responsibility of goods, materials and merchandise must be provided the means to properly discharge that responsibility. Warehouses and other storage space must be equipped with adequate physical protection to secure the goods stored within. Authorizations to enter such storage areas must be strictly limited, and the responsible employee must have means to further restrict access in situations where he may feel that the security of the goods under his control is endangered.

Receiving clerks must have adequate facilities for storage or supervision of goods received until they can be passed on for storage or for other use. Shipping clerks must also have the ability to secure goods in dock areas until they are received and loaded by truckers. Without the proper means of securing merchandise during every phase of its handling, assigned personnel cannot be held responsible for merchandise intended to be under their control, and the entire system will break down. Unreasonable demands, such as requiring a shipping clerk to handle the movement of merchandise in such a way that he is required to leave unprotected goods on the dock while he fills out the rest of the order, lead to the very reasonable refusal of personnel to assume responsibility for such merchandise. And when responsibility cannot be fixed, embezzlement is sure to result.

Other Areas

The Mailroom

The mailroom can be a rich field for a company thief to mine. Not only can it be used to mail out company property to an ally or to a set-up address, but it deals in stamps—and stamps are money. Any office with a heavy mailing operation must conduct regular audits of the mailroom.

Many firms have taken the view that the mailroom represents such a small exposure that close supervision is unnecessary. The head of the mailroom in a fair-sized Eastern firm got away with over $100,000 in less than three years from his manipulation of the postal meter. Only a firm that can afford to lose $100,000 in less than three years should think of its mailroom as inconsequential in its security plan.

Trash Removal

Trash removal has presented many problems. Employees have hidden office equipment or merchandise in trash cans and have then picked up the loot far from the premises in cooperation with the driver of the trash pick-up vehicle. Some firms have had a problem when they put out trash on the loading dock to facilitate pick-up. Trash collectors made their calls during the day and often picked up unattended merchandise along with the trash. On-premises trash compaction is one way to end the use of trash containers as a safe and convenient vehicle for removing loot from the premises.

Every firm has areas which are vulnerable to attack—what and where they are can only be determined by thorough surveys and regular re-evaluation of the entire operation. There are no short cuts. The important thing is to locate the areas of risk and set up procedures to reduce or eliminate them.

WHEN CONTROLS FAIL

Investigation

There are occasions when a company is so beset by internal theft that problems seem to have gotten totally out of hand. In such cases it is often difficult to localize the problem sufficiently to set up specific countermeasures in those areas affected. The company seems simply to "come up short." Management is at a loss to identify the weak link in its security, much less how theft is accomplished after security has been compromised.

Undercover Agents

In such cases, many firms similarly at a loss, in every sense of the word, have found it advisable to engage the services of security firm which can provide undercover agents to infiltrate the organization and observe the operation from within.

Such an agent may be asked to get into the organization on his own initiative. The fewer people who know of his presence, the greater his protection, and the more likely he is to succeed in his investigation. It is also true that when large scale thefts take place over a period of time, almost anyone in the company could be involved. Even one or more top executives could be involved in serious operations of this kind, therefore secrecy is of great importance. Since several agents may be used in a single investigation, and since they may be required to find employment in the company at various levels, they must have, or very convincingly seem to have, proper qualifications for the level of employment they are seeking. Over- or under-qualification in pursuit of a specific area of employment can be a problem, so they must plan their entry carefully. Several agents may have to apply for the same job before one is accepted.

Having gotten into the firm's employ, the agent is on his own. He must conduct his investigation and make reports to his office with the greatest discretion to avoid discovery. But he is in the best possible position to get to the center of the problem, and such agents have been successful in a number of cases of internal theft in the past.

These investigators are not inexpensive, but they earn their fee many times over in breaking up a clever ring of thieves.

It is important to remember, however, that these men are trained professionals. Most of them have had years of experience in undercover work of this type. Under no circumstances should a manager think of saving money by using employees or well-meaning amateurs for this work. Such a practice could be dangerous to the inexperienced investigator and would almost certainly warn the thieves, who would simply withdraw from their illegal operation temporarily until things had cooled down, after which they could return to the business of theft.

Prosecution

Every firm has been faced with the problem of establishing policy regarding the disposal of a case involving proven or admitted employee theft. They are faced with three alternatives: to prosecute, to discharge, or to retain the thief as an employee. The policy they have established has always been difficult to arrive at, because there is no ready answer. There are many proponents of each alternative as the solution to problems of internal theft.

However difficult it may be, every firm must establish a policy governing matters of this kind. And the decision as to that policy must be arrived at with a view to the greatest benefits to the employees, the company, and to society as a whole. An enlightened management would also consider the position of the as-yet-to-be-discovered thief in establishing such policy.

Discharging The Thief

Most firms have found that discharge of the offender is the simplest solution. Experts estimate that 90 percent of those employees discovered stealing are simply dismissed; most of those are carried in the company records as having been discharged for "inefficiency" or "failure to perform duties adequately."

This policy is defended on many grounds, but the most common are:

- Discharge is a severe punishment and the offender will learn from his punishment.

- Prosecution is expensive.

- Prosecution would create an unfavorable public relations atmosphere for us.

- If we reinstate him in the company—no matter what conditions we place on his reinstatement—we will appear to condone theft.

- If we prosecute and he is found not guilty, we will be open to civil action for false arrest, slander, libel, defamation of character and other damages.

There is some validity in all of these views, but each one bears some closer scrutiny.

As to learning (and presumably reforming) as a result of discharge, experience does not bear out this contention. In a recent study, a security organization found that 80 percent of the known employee thieves they questioned with polygraph substantiation admitted to thefts from previous employers. Now it might well be argued that, since they had not been caught and hence discharged as a result of these prior thefts, the proposition that discharge can be therapeutic still holds, or at least has not been refuted. That may be true and it should be considered.

Prosecution is unquestionably expensive. Personnel called as witnesses may spend days appearing in court. Additional funds may be expended investigating and establishing a case against the accused. Legal fees may be involved. But can a company afford to appear so indifferent to significant theft that it refuses to take strong action when it occurs?

As to public relations, many experienced managers have found that they have not suffered any decline in esteem. On the contrary, in cases where they have taken strong, positive action, they have been applauded by employees and public alike. This is not always the case, but apparently a positive reaction is usually the result of vigorous prosecution in the wake of substantial theft.

Reinstatement is sometimes justified by the circumstances. There is always, of course, a real danger of adverse reaction by the employees, but if reinstatement is to a position not vulnerable to theft, the message may get across. This is a most delicate matter that can be determined only on the scene.

As far as civil action is concerned, that possibility must be discussed with counsel. In any event, it is to be hoped that no responsible businessman would decide to prosecute unless the case was a very strong one.

Borderline Cases

Even beyond the difficulty of arriving at a satisfactory policy governing the disposition of cases involving employee theft, there are the cases which are particularly hard to adjudicate. Most of these involve the pilferer, the long-time employee, or the obviously upright employee who finds himself in financial difficulty and steals out of desperation. In each case the offender freely admits his guilt and pleads that he was overcome by the temptation suddenly thrust before him.

What should be done in such cases? Many companies continue to employ such men, provided they make restitution. They are often found to be grateful, and they continue to be effective in their jobs.

In the last analysis it is each individual businessman who must make the determination of policy in these matters. Only he can determine the mix of toughness and compassion that will guide the application of policy throughout.

Hopefully, every manager will determine to avoid the decision by making employee theft so difficult—so unthinkable—that it will never occur. That goal may never be reached, but it's a goal to strive for.

Exam Up To Here.

Chapter 4

BASICS OF
DEFENSE

In order to set up defenses against losses from crime, accident or natural disasters, there must first be some means of identification and evaluation of the risks from which action can be initiated to reduce or eliminate them.

THE SECURITY SURVEY

The identification of the risks can only be accomplished by an exhaustive physical examination of the premises and a thorough inspection of all operational systems and procedures. Special attention should be devoted to those procedures involving the handling, receiving, shipping and storage of goods, materials and supplies; and the flow of documents authorizing expenditures of any kind, or dealing with the movement or inventory of merchandise and other assets.

Such an examination or survey has as its overall objective the analysis of a facility to determine the existing state of its security; to locate weaknesses in its defenses; to determine the kind and degree of protection required; and ultimately to lead to recommendations establishing a total security program.

Motivation setting the survey in motion should come from top management in order to insure that adequate funds for the undertaking are available and to guarantee the cooperation of all personnel in the facility. Since a thorough survey will require an examination of procedures and routines in regular operation as well as an inspection of the physical plant and its environs, management's interest in the project is of the highest priority.

The survey may be conducted by staff security personnel or by qualified security specialists employed for this purpose. Some experts suggest that outside security people could approach the job with more objectivity and would have less of a tendency to take certain areas or practices for granted, thus providing a more complete appraisal of existing conditions.

Whoever undertakes the survey, it is important that they have training in the field and that they have achieved a high level of ability. It is also important that at least some members of the survey team be totally familiar with the facility and its operation. Without such familiarity it would be difficult to formulate the survey plan, and the survey itself must be planned in advance in order to make the best use of personnel and in order to study the operation in every phase.

Part of the plan may come from previous studies and recommendations; these should be studied for any information they may offer. Another part of the survey plan will include a check list made up by the survey team in preparation for the actual inspection. This list will serve as a guide and a reminder of areas that must be examined and, once drawn, should be followed systematically. In the event some area or procedure has been omitted in the preparation of the original check list, it should be included in the inspection and its disposition noted in the evaluation and recommendation.

Since no two facilities are alike—not even those in the same business—no check list exists that could universally apply for survey purposes. The following discussion is intended only to indicate those areas where a risk may exist. It can be considered as merely a guide to the kinds of questions or specific problems that might be dealt with.

The Facility

- Consider the *perimeter* as a security problem. Check fencing, gates, culverts, drains, etc. Check lighting—including standby lights and power. Check overhangs and concealing areas. Can vehicles drive up to fence?

- Consider the *parking lot* as a problem. Are employee automobiles adequately protected from theft or vandalism? How? Is the lot sufficiently isolated from plant or office to prevent unsupervised back and forth traffic? Are there gates or turnstiles for the inspection of traffic if that is necessary? Are these inspection points properly lighted? Can packages be thrown over or pushed through the fence into or out of the parking lot?

- Consider all *adjacent building* windows and roof tops as security problems. Are spaces near these adjacencies accessible to them? Are they properly secured? How?

- Consider all *doors and windows* less than 18 feet above ground level as security problems. How are these openings secured?

- Consider the *roof* as a security problem. What means are employed to prevent access to the roof?

- Consider the issuance of *main entrance keys* to all tenants in a building a security problem. How often are entrance locks changed? What is building procedure when keys are lost or not returned? How many tenants are in the building? What business are they in?

- Consider any *shared occupancy*—as in office buildings—a security problem. Does the building have a properly supervised sign-in log for off-hours? Do elevators switch to manual, or can floors be locked against access outside of business hours? When are they so switched? By whom can they then be operated? Who collects the trash, and how and when is it removed from the building? Are lobbies and hallways adequately lighted? What guard protection does the building have? How can they be reached? Are washrooms open to the public? Are equipment rooms locked? Is a master key system in use? How are keys controlled and secured? Is there a receptionist or guard in the lobby? Can the building be accessed by stair or elevator from basement parking facilities?

- Consider all areas containing *valuables* to be a security problem. Do safes, vaults, or rooms containing valuables have adequate alarms? What alarms are in place to protect against burglary, fire, robbery or surreptitious entry?

- Consider the *off-hours* when the facility is not in operation or all nighttime hours to be a security problem. How many guards are on duty at various times of day? Are guards alert and efficient? How are guards equipped?

How many patrols are there and how often do they make their rounds? What is their tour? What is the guard communication system?

- Consider the control and supervision of *entry into the facility* a security problem. What method is used to identify employees? How are applicants screened before they are employed? How are visitors (including salesmen, vendors and customers) controlled? How are privately owned vehicles controlled? Who delivers the morning mail and when? How are empty mail sacks handled? Do you authorize salesmen or solicitors for charity in the facility? How are they controlled? Are their credentials checked? Who does the cleaning? Do they have keys? Who is responsible for these keys? Are they bonded? Who does maintenance or service work? Are their tool boxes inspected when they leave? Are their credentials checked? By whom? Are alarm and telephone company men allowed unlimited access? Is the call for their service verified? By whom? How is furniture or equipment moved in or out? What security is provided when this takes place at night or on weekends? Are messengers permitted to deliver directly to the addressee? How are they controlled? Which areas have the heaviest traffic? Are visitors claiming official status, such as building or fire inspectors, permitted free access? Are their credentials checked? By Whom?

- Consider *keys and key control* a security problem. Are keys properly secured when not in use? Are locks replaced or recored when a key is lost? Are locks and locking devices adequate for their purpose? Are all keys accounted for and logged? What system is used for the control of master and sub-master keys?

- Consider *fire* a basic security problem. Are there sufficient fire boxes throughout the facility? Are they properly located? Is the type and number of fire extinguishers adequate? Are they frequently inspected? How far is the nearest public fire department? Have they ever been invited to inspect the facility? Does the building have automatic sprinklers and automatic fire alarms? Are there adequate fire barriers in the building? Is there an employee fire brigade? Are fire doors adequate? Are "No Smoking" signs enforced?   Are flammable substances properly stored?   Is there a program of fire prevention education?   Are fire drills conducted on a regular basis?

For all its seeming length, this is in reality only a sample of the kinds of questions that must be asked in conducting a survey of any facility. This list covers only a general overview of some of the aspects to be covered.

Departmental Evaluations

Each department in the organization should be evaluated separately in terms of its potential for loss. These departmental evaluations will eventually be consolidated into the master survey for final recommendation and action. Basic questions might be as follows:

· Is the department function such that it is vulnerable to embezzlement?

· Does the department have cash funds or negotiable instruments on hand?

· Does the department house confidential records?

· What equipment, tools, supplies or merchandise can be stolen from the department?

· Does the department have heavy external traffic? Internal traffic?

· Does the department have "target" items in it such as drugs, jewelry, furs, etc.?

· What is the special fire hazard in the department or from adjacencies?

These are questions that may serve to guide the survey in focusing on particular areas of risk in each department to be examined. Where particular risks predominate, special attention must be paid to provide some counter-action to remove them.

Personnel Department

Can the department area be locked off from the rest of the floor or building after hours?

- How are door and file keys secured?
- Are files kept locked during the day when not in use?
- What system is followed with regard to the payroll department when employees are hired or terminated ?
- What are the relationships between persons in personnel and payroll?
- What are the employment procedures? How are applicants screened?
- How closely does personnel work with security on personnel employment procedures?

Security of personnel files is of extreme importance. Normally these files will contain information on every employee, past and present, from the president on down. This information is highly confidential and must be handled that way. There can be no exceptions to this firm policy.

Accounting

The accounting department has total supervision over the firm's money and will generally be the area most vulnerable to major loss due to crime. Certainly protective systems have been in operation in this area from the company's founding, but these systems must be re-evaluated regularly, in light of ongoing experience, to find ways of improving both their efficiency and security.

Cashier

- How accessible is the cash operation to hallways, stairs and elevators?
- Do posted signs clearly announce the location and operating hours of the cashier?
- Is there generally sufficient cash on hand to invite an employee to abscond with it? To attract an attempted burglary or armed robbery?
- What are the present systems of audit and controls? What forms are used?
- What are the controls put on cashier embezzlement ?
- What are the opportunities for collusion in this operation?

Is the security adequate to the risk?

Accounts Receivable

In evaluating the frauds to which accounts receivable is vulnerable, it will be necessary to consider, from experience, all the possibilities, and in this light to examine every step of the procedures currently in operation, from billing advice through billing to the credit of the account. All flow of information and action documents must be studied minutely to find if any flaw or weakness could be exploited for criminal purposes.

- Consider the billing procedure with particular attention to the forms used and the authorizations required.
- Try to determine how difficult it might be to cash a check payable to the company.
- What are the opportunities to destroy billing records and to keep and cash a check?
- What are the possibilities of altering invoices to show a lesser amount payable?

Accounts Payable

Accounts payable as a disbursing entity invites more attention from thieves than most other areas. It is particularly susceptible to internal attack in a number of ways. The most common is the dummy invoice by which forged authorizations permit payment to a nonexistent account. This is relatively easy to overcome, however, by an alert staff working within a system that provides reasonable security.

- Examine all forms and systems on a step-by-step basis.
- How is the authenticity of new accounts established?

Payroll

- What system is used to introduce a new employee into the payroll?
- Do the records in personnel and in payroll correspond? Are these cross checked? How? By whom?
- Could an employee in personnel conspire with an employee in payroll to introduce fictitious employees into the records? What would prevent this?

Company Bank Accounts

- Can one person transfer unlimited funds?
- Is there a ceiling limiting withdrawal of company funds?
- What instructions have been given to the bank? By whom? Who can change these instructions?
- Who audits company bank accounts? How often?
- Could the treasurer or controller leave with the company bank account?

It is well to remember here, especially when it appears that questions concerning the probity of company officers are posed, that the job of the security survey is not to make judgments on whether a criminal act is *likely* to occur, but whether it *could.* A survey of this nature in no way implies that the treasurer is apt or in any way inclined to abscond with company funds. It simply asks whether or not procedures are such that he could. If this were the case, the recommendation would surely suggest that safeguards such as countersignatures be set up so that the current or future treasurers could not perform what they very likely would not perform in any event.

Data Processing

- Are adequate auditing procedures in effect on all programs?
- How are printouts of confidential information handled?
- What is the off-site storage procedure? How are such tapes updated?
- What is the system governing program access?
- How is computer use logged? How is the accuracy of this record verified?
- Who has keys to computer spaces? How often is the list of authorized key holders evaluated?
- What controls are exercised over access? How often is the list of those authorized to enter updated?
- What fire prevention and fire protection procedures are in effect? What training is given employees in fire prevention and protection? What is the number, location

and condition of fire extinguishers and the basic extinguishing system?

Purchasing

This is an area subject to many temptations. Graft is often freely offered in the form of cash, expensive gifts, lavish entertainment and luxurious vacations—all in the name of seeking the good will of the purchasing agent. Generally speaking, this is not a security matter unless he succumbs to the extent that he is paying for goods never delivered or perhaps paying an invoice twice. If his judgment is distorted by all the attention from vendors and he buys unwisely, that is a concern of management in which security plays no role.

There are, however, some areas in the purchasing function in which security might be involved:

- What are the procedures preventing double payment of invoices? Fraudulent invoices? Invoices for goods never received? How often is this area audited?

- Are competitive bids invited for all purchases? Must the lowest bid be awarded the contract?

- What forms are used for ordering, authorizing payment, etc.? How are they routed?

- Since purchasing is frequently responsible for the sale of scrap, waste paper and other recoverable items, who verifies the amount actually trucked away? Who negotiates the sale of waste or scrap material? Are several prospective buyers invited to bid? How is old equipment or furniture sold? What records are kept of such sales? Is the system audited? What controls are placed on the authority of the seller?

Shipping and Receiving Areas

Freight and merchandise handling areas are particularly troublesome. There is a great potential for theft in these areas. Close attention must be paid to current operations and efforts must continually be directed to their improvement.

- What inspections are made of employees entering or leaving such areas?

- How is traffic in and to such areas controlled? Are these areas separated from the rest of the facility by a fence or barrier?

- Where is merchandise stored after receipt or before ship-ment? What is the security of such area? What is the nature of supervision in these areas?
- What is the system for accountability of shipments and receipts?
- Is the area guarded?
- What losses are being experienced in these areas? What is the profile of such loss (type, average amount, time of day, etc.)?
- Is merchandise left unattended in these areas?
- Are truck drivers provided with rest room facilities separate from dock personnel? Are they isolated from them at all times to prevent collusion?
- How many people and who are authorized in security storage areas?

Miscellaneous
- What records are kept of postage meter usage?   What controls are established over meter usage?
- How is the use of supplies and materials controlled ?
- How are forms controlled?

Report of the Survey

After the survey has documented the full scope of its exami-nation, a report should be prepared indicating those areas which are weak in security and recommending those measures which might reasonably bring the security of the facility up to acceptable stand-ards. On the basis of the status as described in the survey, and considering the recommendations made, the security plan can now be drawn up.

In some cases compromises may have to be made. The siting of the facility or the area involved, for example, may make the ideal security program of full coverage of all contingencies too costly to be practical. In such cases the plan must be re-studied to find the best approach to achieve acceptable security standards within these limitations.

It must be understood that the security director will rarely get all of what he wants to do the job. As in every department, he must work within the framework of the possible. Where he is denied extra personnel, he must find hardware that will help to replace them. Where his request for more coverage by CCTV is turned down, he must develop inspection procedures or barriers that may serve a similar purpose. If, at any point, he feels that he has been cut to a point where the stated objective cannot be achieved, he is obliged to communicate that opinion to management, who will then determine whether to diminish their original objective or authorize more money. It is important, however, that the security director exhaust every alternative method of coverage before he goes to management with an opinion that requires this kind of decision.

Periodic Review

Even after the security plan is formulated, it is essential that the survey process be continued. A security plan, to be effective, must be dynamic. It must change regularly in various details to accommodate changing circumstances in a given facility. Only regular inspections can provide a feel for the needs of the operation, and only periodic surveys can provide a basis for the ongoing evaluation of the security status of the company. Exposures and vulnerability change constantly. What may appear to be a minor alteration in operational routines may have a profound effect on the security of the entire facility.

Security Files

The survey and its resultant report are also valuable in the building of security files. From this evaluation emerges a detailed current profile of the firm's regular activities. With such a file, the security department can operate with increased effectiveness, but it should, by inspections and additional surveys, be kept current.

Such a data base could be augmented by texts, periodicals, official papers and relevant articles in the general press related to security matters. Special attention should be paid to subjects of local significance. Although national crime statistics are significant and help to build familiarity with a complex subject, local conditions have a more immediate import to the security of the company.

As these files are broadened, they will become increasingly useful to the security operation. Patterns may emerge, seasons may become significant, economic conditions may predict events to be alert to.

For example:

- Certain days or seasons may emerge as those on which problems occur.
- Targets for crime may become evident as more data is amassed. This may enable the security director to re-assign priorities.
- A profile of the type and incidence of crimes—possibly even the criminal himself—may emerge.
- Patterns of crime and its modus operandi on payday or holiday weekends may become evident.
- Criminal assaults on company property may take a definable or predictable shape or description, again enabling the security director to better shape his counter-measures.

The careful collection and use of data concerning crime in a given facility can be an invaluable tool for the conscientious security man. It can add an important dimension to his regular re-examination of the status of crime in his company.

SECURITY PERSONNEL

The evaluation of the security survey will ultimately result in a determination of the degree of security required in all areas. Decisions must also be made as to the means by which such security can be most efficiently, effectively and economically achieved. New policies and procedures may be formulated, physical aids to security may be ordered, and the size and deployment of security personnel will be determined. All of these factors must be balanced in the consideration of the protection of the facility to arrive at a formula providing the most protection at the least expense.

How Many?

The need for security personnel is generally proportional to the size of the facility, expressed both in terms of square footage or acreage and number of employees involved. Small businesses of twenty or thirty employees rarely require, and even more rarely can afford, the luxury of a security service of any size. At some point, however, as we consider larger and larger facilities, there is a need for such personnel. This can only be determined by the individual needs of a particular firm as demonstrated by a survey, and as further permitted by the funds available.

Where the security needs of a firm indicate the use of security personnel, they can be the single most important element in the security program. Since they are also the largest single item of expense, they must be used with the greatest efficiency possible. Only careful individual analysis of the needs of each facility will determine the optimum number and use of security personnel. For example, premises with inadequate perimeter barriers would need a larger security force than one with an effective barrier. In determining security needs, therefore, it is important that all protective elements be considered as a supportive whole.

Contract vs. Proprietary Personnel

Assuming that the survey indicates the need for security personnel, the company can supply its needs from a guard service that supplies such personnel on a contract basis, or it can set up its own operation.

Generally speaking, a company can effect substantial savings by contracting with a guard service. It is not obligated for the capital outlay for uniforms or equipment, it has no expense in training or indoctrination, and it is spared the administrative difficulties involved in scheduling, sick leave, absenteeism and salary disputes. Supervision of such personnel is furnished by the contract security company on a full-time basis. The supervision by the security service does not, of course, preclude overall supervision by a small staff of in-house security personnel if such seems indicated. Some firms have turned the entire problem of security—from alarms, surveys and systems to guards and inspections—over to a contract security company.

Qualifications and Training

Whichever kind of force seems indicated, it is essential that security personnel meet high standards of character and loyalty. They must be in good enough physical condition to undergo even arduous exertion in the performance of their duties. They must have adequate eyesight and hearing and have full and effective use of their limbs. In some circumstances exceptions may be made, but these would be for assignments to posts requiring little or no physical exertion or dexterity. They must be of stable character and should be capable of good judgement and resourcefulness.

All applicants for such positions must be carefully investigated. Since they will frequently handle confidential material as well as items of value and will, in general, occupy positions of great trust, they must be of the highest character. Each applicant should be fingerprinted and checked through local and federal agencies. A background investigation of his habits and associates should also be conducted. Signs of instability or patterns of irresponsibility should disqualify him.

Training

Development of training of security personnel must be a continuing concern of management. Development in security is largely accomplished by a thorough training program to improve the security officer's skills and knowledge and to keep him current with the field. The merits of training will be reflected in his attitude, better morale and increased incentive. Training provides increased opportunities for promotion and a better understanding of his relationship to management and the objectives of his job.

Above all, never believe that a former law officer does not need training. He does. In order for him to be successful in security, he must develop new skills and forget some of his previous training.

A Training Program

A training program would cover a wide variety of subjects and procedures. Among them would be the following:

- Company orientation and indoctrination .
- Company and security department policies, systems and procedures.
- Operation of each department .

- Background in applicable law (citizen's arrest, search and seizure, individual rights, rules of evidence, etc.).
- Report writing.
- General and special orders.
- Discipline.
- Self defense.
- First aid.
- Pass and identification systems.
- Package and vehicle search.
- Communications procedures.
- Techniques of observation.
- Operation of equipment.
- Professional standards, including attitudes toward employees.

Security personnel must be cheerful, cooperative and tolerant in their dealings within the company. They must be patient. They must understand that they are not members of law enforcement, but employees (contract or proprietary) who have the job of providing security. If they are good-humored, tactful, patient, and professional, their fellow employees will learn to respect them and look to them for assistance in many ways—frequently beyond the scope of their duties. They should be encouraged to provide whatever assistance they can, since this can increase employee respect for them as men as well as professional security people, thereby encouraging cooperation. And employee cooperation in the security field is vital to its success.

Duties of Security Personnel

The duties of the security officer are many and varied, but among them are common elements that can serve as a guide to every security manager.

1) He protects the buildings and grounds to which he is assigned, including the contents, occupants and visitors.

2)   He enforces rules and regulations governing the facility.

3)   He directs traffic, both foot and vehicular.

4)   He maintains order on his post and helps those persons requiring assistance or information.

5)   He familiarizes himself with all special and general orders, and carries them out to the letter.

6)   He supervises and enforces applicable systems of identifying personnel and vehicles, conducts package and vehicle inspection, and apprehends those persons entering or leaving the facility without required authorization.

7)   He conducts periodic prescribed inspections of all areas at designated times to ascertain their condition of security and safety.

8)   He acts for management in maintaining order and reports any incidents which disrupt such order.

9)   He reports incidents of employees engaged in horseplay, loitering or in violation of clearly stated policy. He reports all sickness or accidents involving employees.

10)   He instantly sounds the alarm and responds to fires.

11)   He logs and turns in any lost or unclaimed property. In the event any property is reported stolen, he checks the recovered property log first before proceeding in the matter.

12)   He makes full reports to his supervisor on all unusual circumstances.

Posts and Patrols

Security personnel may be assigned to a variety of posts, but these fall into just a few categories. They may be assigned to a fixed post, to a patrol detail, or to reserve.

Fixed posts may be gate houses, building lobbies, or guarding a particularly sensitive or dangerous location. *Patrol duty* involves walking or riding a given route to observe the condition of the facility. The perimeter is an important patrol, as are warehouse areas, or open yard storage areas. *Reserves* are those persons standing by in the event assistance is needed by security personnel on fixed posts or patrol duty. The scope of their special orders varies from company to company, but a list of those things that might be required will give the flavor of the tour of duty in an industrial facility.

Security personnel on patrol will make their tours on routes or in areas assigned by the supervisor in charge. They must be fully aware of all policies and procedures governing their tour as well as those that govern the area patrolled.

1) Make sure that the area is secure from intrusion and all gates and other entrances as prescribed are closed and locked. In interior spaces they must check to see that all doors, windows, skylights and vents as prescribed are locked and secure against intrusion as well as possible damage from the weather.

2) Turn off lights, fans, heaters and other electrical equipment when its operation is not indicated.

3) Check for unusual conditions, including accumulations of trash or refuse, blocking of fire exits, access to fire-fighting equipment, etc. Any such conditions, if not immediately correctable, must be reported immediately.

4) Check for unusual sounds and investigate their source. Such sounds might indicate an attempted entry, the movement of unauthorized personnel, the malfunctioning of machinery, or any other potentially disruptive problem.

5) Check any unusual odors and report them immediately, if the source is not readily discovered. Such odors frequently indicate leakage or fire.

6) Check for damage to doors, tracks or weight guards. In cases where doors have been held open by wedges, tiebacks or other devices, these should be removed and their presence reported at the end of the tour of duty.

7) Check for running water in all areas, including wash rooms.

8)   Check to see that all fire-fighting equipment is in its proper place and that access to it is in no way obstructed.

9)   Check whether all processes in the area of the patrol are operating as prescribed.

10)   Check the storage of all highly flammable substances such as gasoline, kerosene, volatile cleaning fluids, etc., to assure that they are properly covered and properly secured against ignition.

11)   Check for cigar or cigarette butts. Report the presence of such butts in "no smoking" areas.

12)   Report the discovery of damage or any hazardous conditions, whether or not they can be corrected.

13)   Exercise responsible control over watchman and fire alarm keys and keys to those spaces as may be issued.

14)   Report all conditions which are the result of violations of security or safety policy. Repeated violations of such policies will require investigation and correction.

The Need for Supervision

Security personnel are, in many respects, the most effective security device available. They rarely turn in false alarms. They can react to irregular occurrences. They can follow a thief and arrest him. They can detect and respond. They can prevent accidents and put out fires. In short, they are human—and they can perform as no machine can.

But as humans they are subject to human frailties. Security personnel must be adequately supervised in the performance of their duties. It is important to be sure that policy is followed; that each member of the security force is thoroughly familiar with policy; and that the training and indoctrination program is adequate to communicate all the necessary information to each member of security. Each guard must be disciplined for a violation of policy, while at the same time management must see to it that he now knows the policy so violated.

All of these elements must be regularly reviewed. It must be remembered that well-trained, well-supervised security personnel may be the best possible protection available, but badly selected, badly supervised guards are not an asset at all—and, worse, could themselves be a danger to security. They can, after all, succumb to temptation like any other individual. The opportunities for theft are far greater than those offered to an average employee.

Issuance and use of keys to stockrooms, security storage and other repositories of valuable merchandise or materials must be limited and their use strictly accounted for. Fire protection not otherwise covered by sensors or sprinkler systems is not often a major problem in such spaces, and in the event a fire were to threaten, the door could be broken or a guardhouse key could be used to enter the endangered area.

Although all security personnel have been subjected to a thorough background investigation before assignment and they have been closely observed during the early period of employment, they may still yield to the heavy pressure of temptation and opportunity presented to them when they have free and unlimited access to all of the firm's goods. Although it is not good, sound personnel practice to show distrust or lack of faith in the sincerity of security personnel, neither is it fair or reasonable business practice to subject them to such a test of probity nor to expose company property to such risks.

ALARMS

In order to balance the cost factors in the consideration of any security system, it is necessary to evaluate the security needs and then to determine how that security can or, more importantly, should be provided. Since the employment of security personnel can be costly, methods must be sought to improve their efficient use and to extend the coverage they can reasonably provide.

Protection provided by physical barriers is usually the first area to be stretched to its optimum point before looking for other protective devices. Fences, locks, grilles, vaults, safes and similar means to prevent entry or unauthorized usage are employed to their fullest capacity. Since such methods can only delay intrusion rather than prevent it, security personnel are engaged to inspect the premises thoroughly and frequently enough to interrupt or prevent intrusion within the time span of the deterrent capability of the physical barriers. In order to further protect against entry, should both barrier and guard be circumvented, alarm systems are frequently employed.

Such systems permit more economical usage of security personnel, and they may also substitute for costly construction of barriers. They do not act as a substitute for barriers per se, but they can support barriers of lesser impregnability and expense, and they can warn of movement in areas where barriers are impractical, undesirable, or impossible.

In determining whether a facility actually needs an alarm system, a review of past experience of robbery, burglary or other crimes involving unauthorized entry should be part of the survey to be evaluated in the formulation of the ultimate security plan. Such experience, viewed in relation to national figures and the experience of neighboring occupancies and businesses of like operation, may well serve as a guide to determine the need for alarms.

Kinds of Alarm Protection

Alarm systems provide for three basic types of protection in the security system.

1) *Intrusion alarms* signal the entry of persons into a facility or an area while the system is in operation.

2) *Fire alarms* operate in a number of ways to warn of fire dangers in various stages of development of a fire; or respond protectively by announcing the flow of water in a sprinkler system, indicating either that the sprinklers have been activated by the heat of a fire or that they are malfunctioning.

3) *Special use alarms* warn of a process reaching a dangerous temperature, either too high or too low; warn of the presence of toxic fumes; or warn that a machine is running too fast. Although such alarms are not, strictly speaking, security devices, they may require the immediate reaction of security personnel for remedial action and thus deserve mention at this point.

Alarms do not, in most cases, initiate any counter action. They serve only to alert the world at large or, more usually, specific reactive forces to the fact that a condition exists against which the facility was alarmed.

Alarm systems are of many types, but they all have common elements. Each one has a sensor of some kind which is designed to respond to a certain condition such as the opening of a door, movement within a room, or the rapid rise of heat. Whenever such condition obtains, the sensor activates a circuit. These circuits can be turned to any use that is desired. They can be designed to sound a bell, dial a phone or punch a tape. The capability of the circuit is limited only by the design of the system deemed most effective.

The questions that must be answered in setting up any alarm system are:

1) Who can respond to an alarm fastest and most effectively?

2) What are the costs of such response as opposed to a response of somewhat lesser efficiency?

3) What is the predicted loss factor as between these alternatives?

Alarm Systems

The response systems currently available are:

1) *The Central Station*—This is a facility set up to monitor alarms indicating fire, intrusion and industrial processes. Such facilities are set up in a location as central as possible to a number of clients, all of whom are serviced simultaneously. Upon the sounding of an alarm, a team of security men is dispatched to the scene and the local police or fire department is notified. Depending on the nature of the alarm, on-duty plant or office protection is notified as well. Such a service is as effective as its response time, its alertness to alarms, and the thoroughness of inspection of alarmed premises.

2) *Proprietary System*—This functions in the same way as a central station system except that it is owned by, operated by and located in the facility. Response to all alarms is by the facility's own security or fire personnel.

Since this system is monitored locally, the response time to an alarm is considerably reduced.

3)   *Local Alarm System*—In this case the sensor activates a circuit which in turn activates a horn or siren or even a flashing light located in the immediate vicinity of the alarmed area. Only guards within sound or hearing can respond to such alarms, so their use is restricted to situations where guards are so located that their response is assured.

Such systems are most useful for fire alarm systems, since they can alert personnel to evacuate the endangered area. In such cases the system can also be connected to local fire departments to serve the dual purpose of alerting personnel and the company fire brigade to the danger as well as calling for assistance from public fire-fighting forces.

4)   *Auxiliary System*—In this system installation circuits are led into local police or fire departments by leased telephone lines. The dual responsibility for circuits and the high incidence of false alarms have made this system unpopular with public fire and police personnel. In a growing number of cities, such installations are no longer permitted as a matter of public policy.

5)   *Local Alarm-by-Chance System*—This is a local alarm system in which a bell or siren is sounded with no predictable response. These systems are used in residences or small retail establishments which cannot afford a response system. The hope is that a neighbor or a passing patrol car will react to the alarm and call for police assistance, but such a call is purely a matter of chance.

6)   *Dial Alarm System*—This system is set to dial a predetermined number or numbers when the alarm is activated. The number selected might be the police or the subscriber's home number, or both. When the phone is answered, a recording states that an intrusion is in progress at the location so alarmed. This system is relatively inexpensive to install and operate, but, since it is dependent on general phone circuits, it could fail if lines were busy or if the phone connections were cut.

Alarm Sensors

The selection of the sensor or triggering device is dependent on many factors. The object, space or perimeter to be protected is the first consideration. Beyond that, the incidence of outside noise, movement or interference must be considered before deciding on the type of sensor that will do the best job.

A brief examination of the kinds of devices available will serve as an introduction to a further study of this field.

1) *Electromechanical Devices* are the simplest alarm devices used. They are nothing more than switches which are turned on by some change in their attitude. For example, an electromechanical device in a door or a window, their most common application, would be held in the open or non-contact position by a plunger on a spring when the door or window was closed.

   Opening either of these entrances would release the plunger, which, under the action of the spring, would move forward, engaging the contacts in the device and thus activating the alarm.

   Such devices operate on the principle of breaking the circuit. Since these devices are simply switches in a circuit, they are normally used to cover several windows and doors in a room or along a corridor. Opening any of these entrances opens the circuit and activates the alarm. They are easy to circumvent in most installations by jumping the circuit.

2) *Pressure Devices* are also switches, activated by pressure applied to them. This same principle is in regular use in buildings with automatic door openers. In security applications, they are usually in the form of mats. These are sometimes concealed under carpeting or, when they logically fit in with the existing decor, are placed in a strategic spot without concealment. Wires leading to them would naturally be hidden in some way.

3) *Taut Wire Detectors* are most often used in perimeter defense, where they are stretched along the top of the perimeter barrier in such a position that anyone climb-

ing the fence or wall would almost certainly disturb them. The tension of the wire is scientifically calibrated, and any change in this tension (either more or less) will activate the alarm. These devices may be used in roof access protection or, occasionally, in the closed area inside the perimeter barrier.

4) *Photoelectric Devices* use a beam of light transmitted for as much as 500 feet to a receiver. As long as this beam is directed into the receiver, the circuit is inactive. As soon as this contact is broken, however briefly, the alarm is activated. These devices, too, are used as door openers.

In security applications the beam is modulated so that the device cannot be circumvented by a flashlight or some other light source, as can be done in non-security applications. For greater security, ultra-violet or infrared light is used, although even these can be spotted by an experienced intruder unless an electronic flicker device is incorporated into the device. Obviously the device must be undetectable, since once the beam is located it is an easy matter to step over or crawl under it.

In some applications a single transmitter and receiver installation can be used, even when they are not in a line of sight, by a mirror system reflecting the transmitted beam around corners or to different levels. Such a system is difficult to maintain, however, since the slightest movement of any of the mirrors will disturb the alignment and the system will not operate.

5) *Motion Detection Alarms* operate with the use of radio frequency transmission or with the transmission of ultrasonic waves.

The *radio frequency* motion detector transmits waves throughout the protected area from a transmitting to a receiving antenna. The receiving antenna is set or adjusted to a specific level of emission. Any disturbance of this level by absorption or alteration of the wave pattern will activate the alarm.

The false alarm rate with this device can be high,

since the radio waves will penetrate the walls and respond to motion outside the designated area unless the walls are shielded. Some such devices on the market permit an adjustment whereby the emissions can be tuned in such a way as to cover only a single area without leaking into outside areas, but these require some skill to tune them properly.

The *ultrasonic* motion detector operates in much the same way as the radio frequency unit, except that it consists of a transceiver which both transmits and receives ultrasonic sound waves. One of these units can be used to cover an area, or they may be used in multiples where such use is indicated. They can be adjusted to cover a single, limited area, or broadened to provide area protection.

The alarm is activated when any motion disturbs the pattern of the sound waves. Some units come with special circuits that can distinguish between inconsequential movement, such as flying moths or moving drapes, and an intruder.

Ultrasonic waves do not penetrate walls and are, therefore, unaffected by outside movement. They are not affected by audible noise per se, but such noises can sometimes disturb the wave pattern of the protective ultrasonic transmission and create false alarms.

6)   *Capacitance Alarm Systems,* also referred to as the proximity alarm, are used to protect metal containers of all kinds. This alarm's most common usage is to protect a high security storage area within a fenced enclosure. To set the system in operation, an ungrounded metal object, such as the safe, file or fence mentioned above, is wired up to two oscillator circuits which are set in balance. An electromagnetic field is created around the protected object. Whenever this field is entered, the circuits are thrown out of balance and the alarm is initiated. The electromagnetic field may project several feet from the object, but it can be adjusted to operate only a few inches from it where traffic in the vicinity of the object is such that false alarms would be triggered if the field extended too far.

7) *Sonic Alarm Systems,* which are known variously as
    noise detection alarms, sound alarms or audio alarms,
    operate on the principle that an intruder will make
    enough noise in a protected area to be picked up by
    microphones which will in turn activate an alarm. This
    system has a wide variety of uses, limited only by the
    problems of ambient noise levels in a given area. The
    system consists simply of a microphone set in the pro-
    tected area and connected to an alarm signal and re-
    ceiver. When a noise activates the alarm, a monitoring
    guard turns on his receiver and listens in on the
    prowler.
        Such a system must be carefully adjusted to avoid
    alarming at every noise. Usually adjustment is turned
    to alarm at sounds above the general level common
    to the protected area. The system is not useful in areas
    where background noise levels are so high that they
    will drown out the anticipated sounds of surreptitious
    entry. This device, too, may come with a sound dis-
    criminator which evaluates sounds to eliminate false
    alarms.

8) *Vibration Detectors* provide a high level of protection
    against attack in specific areas or upon specific objects.
    In this system a specialized type of contact micro-
    phone is attached to an object, such as works of art,
    safes, files, etc., or to a surface such as a wall or a ceil-
    ing. Any attack upon or movement of these objects
    or surfaces causes some vibration. This vibration is
    picked up by the microphone, which in turn activates
    an alarm. The unit may be adjusted as to sensitivity,
    which will be set according to its application and its
    environment. Here again, a discriminator unit is avail-
    able to screen out harmless vibrations. This unit is
    very useful in specific application, since its false alarm
    rate is very low.

Other Perimeter Alarm Devices
Certain additional alarm devices are currently in use as perimeter
protection. These are:

- *The Electronic Fence,* which consists of from three to nine wires stretched along a cleared space inside the perimeter barrier. These wires are spaced well apart and serve as antennae which, when acting to conduct an electric circuit, create an electromagnetic field as in the capacitance system. Any approach to this fence disturbs the balance of the field and initiates an alarm. It is important that this fence be protected from false alarms, which could be triggered by animals, tall grass or any other casual and innocent disturbance.

- *The Radio Frequency System* is a system in very limited use, and largely considered as experimental. This system operates somewhat like a photoelectric system, except that radio waves are used in a narrow beam projected from a transmitter to an antenna. Any disturbance of energy thus transmitted activates the alarm.

Cost Considerations

The costs involved in setting up even a fairly simple alarm system can be substantial, and the great part of this outlay is nonrecoverable should the system prove to be inadequate or unwarranted. Its installation should be predicated on exposure, concomittant need, and the manner of its integration into the existing or planned security program.

This last is important to consider, because the effectiveness of any alarm procedure lies in the response it commands. As elementary as it may sound, it is worth repeating that an alarm takes no action; it only notifies us that action should be taken. There must, therefore, be some entity near at hand that can take that action. Too often, otherwise effective alarm systems are set up without adequate supportive or responding personnel. This is at best wasteful and at worst dangerous.

In many instances alarm installations are made in order to reduce the size of the guard force. This is sometimes possible. At least, if the current guard force cannot be reduced in number, those additional guards needed to cover areas now alarmed will no longer be required.

Good business practice demands that the expense of alarm installations be undertaken only after a carefully considered cost and effectiveness analysis of all the elements. If the existing force of security personnel can cover the security requirements of the facility, no alarm system is needed. If they can cover the ground but not in a way that will satisfy security standards, then more guards will be needed, or the existing force must be augmented by an alarm system that will effectively extend their coverage and their protective ability. The costs and effectiveness must be studied together with an eye toward the efficient achievement of stated objectives.

FIRE ALARM SYSTEMS

Fire alarm systems, like intrusion alarms, can be viewed as consisting of the signaling device and the sensor. The sensor warns of fire or its danger by activating a circuit, whereas the signaling device notifies those concerned of the danger.

Signals

The signaling system is activated by a sensor, a water-flow switch, or a manual alarm. The signal so activated is transmitted to an alarm receiver at a monitoring panel, which activates appropriate alarms and at the same time notifies fire protection personnel of the location of the alarm source. This notification is normally performed through a printer or some other read-out device that records time and location of the alarm. In many areas the alarm, when sounded, will be in such a series of siren sounds or bell strokes that the location of the fire is identified.

The manual fire alarm stations are of two different types:

1) *Local Alarms,* which are clearly identified at the pull box as such. This alarm is designed to alert personnel in the building, or in the area where such alarm sounds, that a fire (or in some cases a drill) is in progress. Depending on the prescribed procedures, personnel so alerted will stand by for further instructions or immediately evacuate the building. It is important to remember, however, that such alarms sound locally only. They do not notify anyone beyond the sound of the alarm of the fire threat. Unless someone should happen to notify the fire department by phone, there will be no response to the alarm.

2)   *Remote Signal Stations* transmit a signal directly to the monitoring panel. This performs manually essentially what the automatic signaling devices do automatically. Normally a local signal is sounded at the same time the signal is being transmitted to the receiver.

Sensors

There are several types of fire sensors, all of which are useful in fire protection systems. Since each performs a different function at different stages of a fire's development, each one has a place in an effective fire protection system.

1)   *The Ionization Detector* operates when exposed to those invisible products of combustion which are emitted by a fire in its earliest stage of development. Since a fire may be in progress for as much as two or three hours before it gives any otherwise detectable evidence of its presence, this sensor is extremely useful in giving early warning of what might be termed an incipient fire. Invisible products of combustion—largely hydrocarbons—interfere with a current passing between two plates in this device and activate an alarm.

2)   *The Photoelectric Smoke Detector* operates on the basic principle of every photoelectric device, in which a beam of light is transmitted to a cell which is in balance as long as the light source is constant. When that source is in any way interrupted, the balance is disturbed and the unit alarms. These detectors will alarm on a concentration of smoke of from 2 to 4 percent.

3)   *The Infrared Flame Detector,* as its name suggests, reacts to the infrared emissions from flame.

4)   *Thermal Detectors* sense the temperature in the protected area. Some are set to alarm when the temperature reaches a fixed level; others respond to a rapid climb in the temperature. This latter type, known as "rate-of-rise" detectors, cannot be used in industrial areas where the temperature fluctuates strongly.

Some detectors combine the features of both the fixed temperature and rate of rise units and are generally considered to be more useful than either unit alone.

Automatic Sprinkler Systems

A second level of alarm is activated by the flow of water in an automatic sprinkler system. Such an alarm is important in that areas covered by sprinkler systems are rarely otherwise alarmed. When the heat level in such areas rises sufficiently, the sprinkler system is activated by the melting of a metal seal and water flows through the system upon the release of sprinkler head valves. This activates an alarm which indicates the flow of water.

LOCKS, KEYS AND SAFES

Locks

Locks are, generally speaking, the cheapest security investment that can be made. Cost-cutting in their purchase is usually a poor economy, since a lock of poor quality is virtually useless and—effectively—no lock at all.

The elementary but often overlooked fact of locking devices is that, in the first place, they are simply mechanisms that extend the door or window into the wall that holds them. If, therefore, the wall or the door itself is weak or easily destructible, the lock cannot be effective.

In the second place, it must be recognized that any lock will eventually yield to an attack. They must be thought of only as delaying devices. But this delay is of primary importance. The longer an intruder is stalled in an exposed position while he works at gaining entry, the greater are the chances of discovery. Since many types of locks in general use today provide no appreciable delay to even the unskilled prowler, they have no place in security applications.

Kinds of Locks

A brief review of the types of locks in general use, with notations of their characteristics, may serve to familiarize those as yet unacquainted with them with the variety available.

Warded Locks are those found generally in pre-war construction, in which the keyway is open and can be seen through. These are also recognized by the single plate which includes the doorknob and the keyway. The security value of these locks is nil.

Disc Tumbler Locks were designed for the use of the automobile industry and are in general use in car doors today. Because this lock is easy and cheap to manufacture, its use has expanded to other areas such as desks, files and padlocks. The life of these locks is limited because of their soft metal construction. Although these locks provide more security than warded locks, they cannot be considered as very effective. The delay afforded is approximately three minutes.

Pin Tumbler Locks are in wide use in industry as well as in residences. They can be recognized by the keyway, which is irregular in shape, and the key, which is grooved on both sides. Such locks can be master keyed in a number of ways, a feature which recommends them to a wide variety of industrial applications, although the delay factor is ten minutes or less.

Lever Locks are difficult to define in terms of security, since they vary greatly in their effectiveness. The best lever locks are used in safe deposit boxes and are, for all practical purposes, pickproof. The least of these locks are used in desks, lockers and cabinets and are generally less secure than pin tumbler locks. The best of this variety are rarely used in common applications such as doors because they are bulky and expensive.

Combination Locks are difficult to defeat, since they cannot be picked and few experts can so manipulate the device as to discover the combination. Most of these locks have three dials which must be aligned in the proper order before the lock will open. Some such locks may have four dials for greater security. Many also have the capability of changing the combination quickly.

Code-Operated Locks are combination-type locks in that no keys are used. They are opened by pressing a series of numbered buttons in the proper sequence. Some of them are equipped to alarm if the wrong sequence should be pressed. The combination of these locks can be changed readily. These are high security locking devices.

Virtually every lock manufacturer makes some kind of special high security lock which is operated by non-duplicable keys. A reliable locksmith or various manufacturers should be consulted in cases of such needs.

Electromagnetic Locks are devices holding a door closed by magnetism. These are electrical units consisting of the electromagnet and a metal holding plate. When the power is on and the door secured, they will resist a pressure of up to 1000 pounds.

Card Operated Locks are electrical or, more usually, electromagnetic. Coded cards—either notched, embossed, or containing an embedded pattern of copper flocks—are used to operate such locks. These frequently are fitted with a recording device which registers time of use and identity of the user. The cards serving as keys may also serve as company ID cards.

Padlocks

Padlocks should be hardened and strong enough to resist prying. The shackle should be close enough to the body to prevent the insertion of a tool to force it. No lock which will be used for security purposes should have less than five pins in the cylinder.

It is important to establish a procedure requiring that all padlocks be locked at all times even when they are not securing an area. This will prevent the possibility of the lock being replaced by another to which a thief has the key.

The hardware used in conjunction with the padlock is as important as the lock itself. It should be of hardened steel, without accessible screws or rivets, and should be bolted through the door to the inside, preferably through a backing plate. The bolt ends should be burred.

Locking Devices

In the previous list we have considered the types of locks that are generally available. It must be remembered, however, that locks must work in conjunction with other hardware which effects the actual closure. These devices may be fitted with locks of varying degrees of security and themselves provide security to various levels. In a security locking system, both of these factors must be taken into consideration before determining which system will be most effective for specific needs.

Double Cylinder Locking Devices are installed in doors that must be secured from both sides. They require a key to open them from either side. Their most common application is in doors with glass panels which might otherwise be broken to allow an intruder to reach in and open the door from the other side. Such devices cannot be used in interior fire stairwell doors, since firemen break the glass to unlock the door from the inside in this case.

Emergency Exit Locking Devices are panic-bar installations allowing exit without use of a key. This device locks the door against entrance. Since such devices frequently provide an alarm feature that sounds when exit is made, they are fitted with a lock which allows exit without alarming when a key is used.

Recording Devices provide for a printout of door use by time of day and by the key used.

Vertical Throw Devices lock into the jamb vertically instead of the usual horizontal bolt. Some versions lock into both jamb and lintel. A variation of this device is the Police Lock, which consists of a bar angled to a well in the floor. The end of the bar contacting the door is curved so that when it is unlocked it will slide up the door, allowing it to open. When it is locked, it is secured to the door at one end and set in the floor at the other. A door locked in this manner is virtually impossible to force.

Electric Locking Devices are installed in the same manner as other locks. They are activated remotely by an electric current which releases the strike, permitting entrance. Many of these devices provide minimal security, since the engaging mechanism frequently offers no security feature not offered by standard hardware. The electric feature provides a convenient method of opening the door. It does not in itself offer locking security. Since such doors are usually intended for remote operation, they should be fitted with a closing device.

Sequence Locking Devices are designed to insure that all doors covered by the system are locked. The doors must be closed and locked in a predetermined order. No door can be locked until its designated predecessor has been. Exit is made through the final door in the sequence, and entry can be made only through that same door.

Door Jambs

Since doors, when closed, are as a rule attached to the jamb, it is essential that the jamb be of as strong construction as the door or the lock. Aluminum jambs, for example, are frequently spread by a crowbar or an automobile jack. If they cannot resist such attack, the door can be opened easily. The locking bolt must be at least an inch into the jamb for security and to help prevent spreading. Cylinders should be flush or inset to prevent their being wrenched out or "popped."

Removable Cores

In facilities requiring that a number of keys be issued, the loss or theft of keys is an ever-present possibility. In such situations it might be well to consider removable cores on all locks. These devices are made to be removed if necessary with a core key, allowing a new core to be inserted. Since the core is the lock, this has the effect of rekeying without the necessity of changing the entire device, as would be the case with fixed cylinder mechanisms.

Keys

As previously noted, keys are generally divided into change keys, submaster keys, master keys, and occasionally grand master keys.

Change keys are those keys in a system which will operate one designated lock.

Submasters are those keys which will operate all locks within a grouping as, for example, all the locks in one building in a complex of buildings.

Master keys will open all the locks in the several groupings involved.

Grand masters are those keys that will open everything in a system involving two or more master key groups. This system is relatively rare, but might be used by a multi-premise operation in which each location was master keyed while the grand master would function on any premise.

Master and grand master keys are normally machined to be very thin so that the use of each one is very limited. This is deemed to be a security measure, guarding against their extensive use in the event of loss or theft. This is a dubious proposition at best, since the loss of one of these keys effectively compromises the system and substantially reduces its security value. Even one or two ventures through the facility with such keys could do serious harm and, thin as they are, they might well stand up for that many uses. Unfortunately, when keys of such sensitivity are lost, rekeying is the only answer.

Key Control

In any sizable facility, rekeying can be very expensive, so it is of extreme importance that all such keys be secured under the strictest supervision and accountability.

There are alternative methods to the disruption and staggering expense that can be involved in rekeying. Outer or perimeter locks can be changed first, and the old locks moved to interior spaces requiring a lower level of security. After an evaluation, a determination of priorities can be made and rekeying can be accomplished over a period of time, rather than requiring one huge capital outlay at once.

Of prime importance is the securing of keys so that such problems do not arise. Master and submaster keys should be issued with great reluctance and then with careful records kept on such issue. At such time as they are not in use, these keys must be secured and inventoried regularly.

All keys issued should be physically inspected periodically to ensure that they have not been lost though unreported as such. It is equally important that keys never be issued to anyone except those demonstrably responsible persons who have compelling need for them. Though possession of keys is frequently a status symbol in many companies, management must never issue them on that basis.

Safes

Safes are expensive, but if they are selected wisely, they can be one of the most important investments in security. It is to be emphasized that safes are not simply safes. They are each designed to perform a particular job to a particular level of protection. To use fire-resistant safes for the storage of valuables—an all too common practice—is to invite disaster. At the same time, it would be equally careless to use a burglary-resistant safe for the storage of valuable papers or records, since, if a fire were to occur, the contents of such a safe would be reduced to ashes.

Ratings

Safes are rated alphabetically to describe the degree of protection they afford. Fire-resistant safes are rated up to "A," which is a safe that will resist an external temperature of up to 2,000° for four hours. Class "B" safes can protect against exposure up to 1850° for two hours; and a Class "C" safe gives one hour of protection up to 1700°. Fire safes are also tested for their resistance to damage in the collapse of a building in a fire. All classifications must meet the same standards in these tests.

Burglary resistance is theoretically rated down from "A," but not until the "E" class safe is there significant protection. This container has a 1" steel body with a 1-1/2" door which is rated as resistant to ripping or cutting with ordinary hand tools.

Any safe rated burglary-resistant by Underwriters Laboratories must include a UL listed combination lock, UL listed relocking device, cast or welded-plate body, and door of special metal alloys which can resist carbide drills. Higher rated safes indicate a rated resistance to torch, tool or explosive attack for varying periods of time—15 minutes for "F," 30 minutes for "H," or 60 minutes for "I."

UL testing for burglary-resistance in safes does not include the use of diamond core drills, thermic lance or other devices yet to be developed by the safecracker.

In some businesses a combination consisting of a fire-resistant safe with a burglary-resistant safe inside may serve, but in no event must the purposes of these two kinds of safe be confused if there is one of each on the premises.

Back-up Protection

Since even the highest rated burglary-resistant safe can be penetrated in time, it must be supported by alarms and, ideally, periodic physical surveillance. A capacitance alarm is the type most generally used to protect a safe. For surveillance a safe should be located where it can be seen. It should be well lighted and perhaps located where it can be seen from the street by every passing patrol car.

All free-standing safes should be secured to the floor to prevent their being hauled away. Setting them in concrete in the floor is another excellent protection which is coming into wide use.

SURVEILLANCE

Surveillance of a facility is normally conducted by patrolling security personnel who watch for any signs of criminal activity. If they spot any, they are in a position to take such action as necessary.

CCTV

Patrols cannot, however, watch everything at once. In order to extend this area of surveillance, closed-circuit television cameras (CCTV) are used to keep corridors, entrances and designated security areas under observation.

Tape recorders are frequently used in conjunction with these cameras for file and reference purposes. Such tape recordings can be invaluable in identifying persons and for reviewing the circumstances of any event.

CCTV cameras available today can be highly effective in almost total darkness. Such installations are expensive, but since one man can sit at a control panel and watch the output of many cameras covering large parts of the facility, the savings in security personnel as well as the added security may make such an investment practical.

Other Cameras

For surveillance after the fact, especially in those cases where a series of thefts has taken place, motion picture or sequence cameras may be used.

Motion picture cameras using high-speed 16mm film and fast lenses can be set up to take pictures in normal light. These cameras can be activated by an alarm or by a switch. The coverage of events is limited by the amount of film in the camera. This is not a totally satisfactory device, since cameras of this kind are never completely silent and need at least normal room light levels to record legibly.

Sequence cameras record still pictures at regular intervals, or they can, by switch control, take a prearranged number of pictures in rapid succession. The time interval between pictures can be adjusted. Pictures can be taken in almost total darkness with infrared sensitive film and an infrared emission.

In either of these methods, cameras should be concealed or at least out of reach and secured.

Both of these methods are considerably cheaper than a CCTV installation, but neither is nearly so effective.

A Word of Caution

There is much equipment available to the security manager today. Some of it is useful, some not, but none of it is better than the use to which it is put or the system into which it is integrated. No equipment stands on its own. It can be useful only if it is employed properly, fully and effectively. In the end it must return, by some kind of estimation, more than the investment that installed it.

Chapter 5

INSURANCE

Many managers still cling to the notion that the most effective means of guarding against unforeseen business losses is insurance, and all too many still use insurance as a substitute for a comprehensive security program. The fallacy in this attitude is twofold.

In the first place, almost all casualty insurance companies have suffered losses in underwriting crime insurance over the last ten years. In fact, payments for insured losses due to crime have regularly exceeded income from premiums since 1963. Obviously no company can continue to operate at a loss; casualty underwriters are no exception.

In the face of the rapid increase in crime against homes, persons and business establishments, most insurance companies have taken drastic steps to counter this trend. They have cancelled or refused to renew policies of insureds who have suffered losses from criminal activity. They have limited allowable coverage to a point well below replacement or even cash value of goods or property. They have limited the extent of coverage. They have set up perimeters which exclude businesses in high crime areas or in high risk enterprises from any coverage at all. And, lastly, they have increased premium rates. Between 1961 and 1971 premium rates for burglary insurance on liquor stores, for example, rose over 200 percent in Los Angeles, 300 percent in Washington, D.C. and 400 percent in New York.

In the second place, it is virtually impossible to insure against all the losses that could be incurred. Hidden damages in loss of company morale, loss of customer confidence, interruption of vigorous participation in a highly competitive market—all are serious if not fatal blows to any business and can never be recompensed.

Clearly, insurance can never be a substitute for a security program. In many cases, the very fact that assets are insured to some degree tends to reduce the interest of the proprietor in instituting reasonable security procedures beyond those minimums specified in his policy. As an aspect of the overall picture, insurance tends also to reduce any interest the insured may have in capturing or prosecuting perpetrators of crimes, thus in effect encouraging the proliferation of like crimes.

Insurance is certainly important. It is clearly necessary for any businessman who wishes to protect himself against loss—to spread the risk—but it must be thought of as supportive rather than as the principal defense against losses due to crime.

INSURANCE AGAINST CRIME

Crime insurance covers the insured in the event of loss from robbery, theft, forgery, burglary, embezzlement and other criminal acts. It is important, however, to know the specific coverage involved in any such policy and the circumstances under which recovery of losses is allowed.

For example, burglary is generally meant to refer to felonious entry and theft by force. In order to collect insurance after such an attack, there must be evidence of forced entry, such as broken locks or windows, tool marks or other clear evidence that burglary was in fact committed. The mere fact that items are missing will in no way establish the fact that the insured has been the victim of a burglar.

Robbery, too, must be specifically established according to contractual definition. Robbery can be loosely defined as the forcible taking of property by violence or the threat of violence aimed at a person or persons covered by the policy. Theft or purse snatching, for example, would not be covered under a provision covering robbery; neither would burglary.

It is, therefore, essential that the terms describing criminal activity be clearly understood so that the nature and the extent of the coverage conforms to the needs of the insured. This is particularly important with companies who are particularly vulnerable to certain kinds of hazards.

It is also extremely important to check policies for exclusionary clauses which may exempt certain crimes from coverage or which simply do not include certain crimes in the contract. This will require a careful examination, since insurance contracts are notoriously long-winded of necessity to cover all of the possible contingencies within the area of their coverage. Certain of the absences of coverage or exclusions can get lost in the sea of verbiage.

Comprehensive Policies

Comprehensive policies covering dishonesty, destruction and disappearance are designed to provide the widest possible coverage in cases of criminal attack of various kinds. The standard form is set up to offer five different kinds of coverage. The insured has the option of selecting any or all of the insuring agreements offered and of specifying the amount of coverage on each one selected. In addition to the coverage options in the standard form, twelve endorsements are also available to the manager having a need for any or all of them.

The coverages available on the standard form consist of:

I —Employee dishonesty bond.

II —Money and securities coverage on the premises.

III—Money and securities coverage off the premises.

IV—Money order and counterfeit paper currency coverage.

V —Depositors' forgery coverage.

Additional endorsements available are:

1)   Incoming check forgery.

2)   Burglary coverage on merchandise.

3)   Paymaster robbery coverage on and off premises.

4)   Paymaster robbery coverage on premises only.

5)   Broad-form payroll on and off premises.

6)   Broad-form on premises only.

7)   Burglary and theft coverage on merchandise.

8)   Forgery of warehouse receipts.

9)   Securities of lessees of safe-deposit box coverage.

10)   Burglary coverage on office equipment.

11)   Theft coverage on office equipment.

12)   Credit card forgery.

Obviously the premium on this coverage will vary according to the number of options selected and the amount of coverage desired for each.

Estimating Coverage

The following chart can prove useful in determining the amount of crime coverage a firm should carry depending upon its exposure.[6] Although this chart cannot specify the kind of coverage needed, since each company will have its own areas of vulnerability, it can indicate the potential dangers managers face, and the coverage recommended to protect against them.

Figure 1: Dishonesty Exposure Index Indicator

1. Total current assets (cash, deposits, securities, receivables, and goods on hand). $_____

   A.   Value of goods on hand (raw materials in process, finished merchandise or products). $_____

   B.   5% of A $_____

   C.   Total current assets less goods on hand. (The difference between 1 and 1A) $_____

   D.   20% of C $_____

2.   Annual gross sales or income. $_____

   A.   10% of 2 $_____

   B.   Total of 1B, 1D and 2A— the firm's dishonesty exposure index. $_____

Figure 2: Suggested Minimum Amounts of Honesty Insurance

Exposure Index		Amount of Coverage		
up to	25,000	$15,000	to	25,000
25,000 to	125,000	25,000		50,000
125,000	250,000	50,000		75,000
250,000	500,000	75,000		100,000
500,000	750,000	100,000		125,000
750,000	1,000,000	125,000		150,000
1,000,000	1,375,000	150,000		175,000
1,375,000	1,750,000	175,000		200,000
1,750,000	2,125,000	200,000		225,000
2,125,000	2,500,000	225,000		250,000
2,500,000	3,325,000	250,000		300,000
3,325,000	4,175,000	300,000		350,000
4,175,000	5,000,000	350,000		400,000
5,000,000	6,075,000	400,000		450,000
6,075,000	7,150,000	450,000		500,000
7,150,000	9,275,000	500,000		600,000
9,275,000	11,425,000	600,000		700,000
11,425,000	15,000,000	700,000		800,000
15,000,000	20,000,000	800,000		900,000
20,000,000	25,000,000	900,000		1,000,000
25,000,000	50,000,000	1,000,000		1,250,000
50,000,000	87,500,000	1,250,000		1,500,000
87,500,000	125,000,000	1,500,000		1,750,000
125,000,000	187,500,000	1,750,000		2,000,000
187,500,000	250,000,000	2,000,000		2,250,000
250,000,000	333,325,000	2,250,000		2,500,000
333,325,000	500,000,000	2,500,000		3,000,000
500,000,000	750,000,000	3,000,000		3,500,000

To determine the suggested minimum amount of insurance required, compute the firm's dishonesty exposure index, and find the recommended coverage by referring to the list under suggested minimum amounts.

Evaluating Risk

However the exposure is calculated, and however much coverage is then deemed necessary, it is essential that they reflect a realistic appraisal of the risk factors actually involved. Every manager dealing with protection factors must ask himself what the risk really is and what would happen to the company if any of the considered potential hazards came to pass.

In the last analysis he is simply spending certain amounts of money on a gamble that certain potentially harmful events will or are likely to occur. In some cases he is betting that, although such events are not likely to occur, they would be so catastrophic if they did that the amount spent on insurance fades into insignificance compared to the possible losses. The balance is a delicate one and requires much thought and expertise to develop the most efficient insurance program.

Insurance rates are, after all, based on actuarial tables which are presumably updated regularly to reflect the experience of various industries or types of business as a whole in losses from various sources or causes. Since most insurance companies are profit-oriented enterprises themselves, the rates also reflect that factor. They must, as part of their profit structure, charge for claims handling, sales commissions, administrative expense, etc. This ultimately means that a manager must evaluate his own risk as against the insurance industry's evaluation of it.

More often than not he will discover he is under-insured, but there are instances when he may well feel that his exposure to loss is well below that of his particular industry as a whole. In such a case, he might properly elect to save on his premium payments by insuring according to his appraisal of his risk (on the basis of the effectiveness of his security program, for example) instead of on the basis of his exposure.

These are never easy decisions to make. The under-insured are risking ruination, while the over-insured are spending substantial sums to no good purpose.

Whatever the evaluation, however, every manager must be thoroughly conversant with the risks he faces and make provisions for them accordingly. His first impulse must be to minimize the risk as much as possible by instituting security devices and procedures that will reduce the possibility of loss or at least will notify him promptly when a loss has taken place. He must then re-evaluate the risks in the light of this tightened or altered security system and then re-evaluate his supportive system of insurance.

Fidelity Bonds

One important kind of insurance which is frequently badly underestimated is the fidelity bond—a form of insurance that will compensate for company losses from dishonest employees. These bonds are issued on the performance of specific, named employees after the bonding company has satisfied itself, through investigation, that such employees do not represent an evident threat to the assets of the company. In effect the bonding company is guaranteeing the insured that bonded employees will perform in good faith—that they will not commit any dishonest acts against their employer. If any of such bonded employees violate this trust, the guarantor—the bonding company—will stand the loss up to the amount insured.

The investigation by the bonding company is valuable in that it provides a further check on the background of employees in sensitive positions as well as underwriting possible losses resulting from a violation of trust.

Amount of Bonding

Most companies do require that employees handling cash or high value merchandise be bonded, but too many of these companies go along on a program calling for five-, ten- or twenty-five thousand dollar bonds, failing to consider that, if bonding is deemed necessary, it must provide for protection against potential damage that such an employee can cause. In situations where there is no system providing a regular, foolproof audit of cash and valuable merchandise, for example, an employee might steal enormous sums over a period of time, even if his daily handle is relatively small.

The Surety Association of America recently published a list of losses from various kinds of businesses caused by bonded employees. They showed the extent of fidelity coverage and the actual loss. The net loss figures are dramatic. A small sample of their much larger list follows.

This sample clearly indicates the problem faced by business today. It also indicates that many businesses are not handling the problem with a coordinated systems approach. It may appear to be hindsight to point out that adequate bonding would have cost these companies the merest fraction of their ultimate losses; yet we can assume that, in most of these cases, a more realistic evaluation of the exposure, risk and insurance costs would have prevented these substantial losses.

Business	Employee	Loss	Bond	Uninsured Loss
Wholesale Produce	Bookkeeper	$185,820	$25,000	$160,820
Retail Dairy	Office Manager	11,000	2,500	8,500
Hospital	Chief Clerk	15,000	5,000	10,000
Machinery Mfr.	Sales Manager	96,940	50,000	46,940
Department Store	Several	81,000	15,000	66,000
Refrigerator Mfr.	Cashier	20,810	5,000	15,810
Rubber Mfr.	Bookkeeper	126,700	26,000	100,700
Advertising	Billing Clerk	90,875	10,000	80,875
Paper Products	Warehouseman	25,551	15,000	10,551

INSURANCE CONSIDERATIONS

Insurance against criminal acts is only one of the many kinds of insurance that must be considered in protecting any business against unforeseen eventualities. The kinds of insurance that are necessary and the amount of coverage in each category will depend upon the nature of the business and the extent of its exposure to various hazards.

In a general way, however, there are certain considerations that must be taken into account before any program can be settled upon. For purposes of this discussion we will omit considerations of liability insurance and confine ourselves to property insurance, which covers structures, goods, and equipment, cash, papers, records and negotiables.

Extent of Coverage

The first consideration in evaluating any kind of coverage must be the nature and extent of the losses covered. These losses could be classified as *direct*—meaning loss or damage to the element concerned; *loss of use*—meaning loss of certain tangible benefits resulting from destruction or damage to the element concerned; or *extra expenses* losses—meaning those costs resulting from loss or damage to the element concerned, such as the rental of office space and/or equipment after a fire has damaged the insured office and equipment.

Loss of Use/Extra Expense Coverage

Most standard policies do not provide loss of use or extra expense coverage. Since both of these matters can represent a very substantial loss to most companies, consideration must be given to expanding the provisions of the coverage to include them. Both of these losses can be covered either by endorsement or by additional policies which will provide that coverage on a broad basis.

Even a small fire in an office may render it inoperable from smoke and water damage, damaged equipment, etc., for a substantial period of time. Even though all the damage is covered and will be cared for, the interim period when revenues may be lost and new facilities temporarily occupied may be as expensive as the fire itself.

Such a situation might be covered by a Business Interruption contract. Here, too, there are options. A Business Interruption policy can be drawn up on a comprehensive basis, which means that it will cover a broad base of situations that might create a stoppage. Such a policy must, of course, be examined for types of incidents which are specifically excluded from coverage. On the other hand, such a contract might be drawn up in which the incidents covered are specified and perhaps limited to just a few potential hazards.

The amount of coverage and the nature of recovery in business interruption contracts can be complicated. If recovery is on an actual loss basis, a careful audit of actual demonstrable losses must be presented to the insuror in order to collect. If the policy is drawn as a valued loss contract, an accountant must certify the daily amount that would be lost if an interruption were to occur. This amount is entered as part of the contract. The premium and recovery are based upon this amount, figured on the specified number of days to be covered for each interruption.

What Is Covered

Every manager must consider the property to be insured and check to make certain that all the property he has designated is covered by the policy issued. Because of the changing nature of casualty insurance in today's market, certain property may be excluded from coverage because of its location, or because the nature of the business creates or subjects it to special hazards that are not included under the basic coverage.

In such cases special policies or endorsements may be required to fill out the insurance program. If the rates for such coverage are deemed excessive, it might be well to reconsider the original insurance program and to provide a sharply limited coverage in these special areas, while at the same time developing special programs to reduce exposure and vulnerability.

Persons Covered

Generally speaking, property insurance is recoverable only by the insured and his agents. This is extended to include heirs named in a will or receivers in a bankruptcy proceeding. If there are others who should be named, such as non-equity lenders or other interested parties, they must be entered into the contract by endorsement to make certain their interests are protected. If such an endorsement is not made, such persons have no rights under the policy and they must seek relief by other means if a loss should occur.

Time Covered

The period of time covered must be checked as well. Most policies are good for a year starting at a specified time of day on the effective date of the contract, and are in force until a specified time of day one year from that date. Some policies are effective for a longer period of time, so it is important to verify the precise period of coverage referred to in the policy.

Incidents Covered

Property insurance may be specific or comprehensive. It may also qualify its coverage in either kind of policy.

Specific policies will name the incident or incidents which are covered and will further specify the degree to which they are covered. For example, in a policy in which the incidence of water damage is covered, there may be some kinds of water damage which are excluded from the coverage offered in a particular contract. It is especially important for a risk manager to be aware of these exclusions and to avoid the common mistake of supposing that, simply because he is covered by a policy that names water damage as a specific covered incident, he is therefore covered for all water damage from every source.

Comprehensive policies cover a wide variety of incidents except for those specifically exempted. Here again they may limit the degree of their coverage of these incidents. They may also specify conditions which will invalidate the provisions of the contract. If, for example, a contract is drawn insuring a building against fire, the contract may state that the provisions of the contract are invalid during any period when the insured permits fire hazards to increase in the insured building.

How Much Insurance?

Since the options are essentially a choice between recovery of the cash value of the property or recovery of its replacement cost, there can be little hesitation in making a decision. Few experts disagree that insuring to the amount of replacement is clearly the wiser course to follow.

Cash Value Coverage

When property is insured for its cash value only, there will almost inevitably be a loss to the insured unless property values decline enough to make up for the extra costs involved in replacement. The latter might include demolition of the remaining structure in the event of fire; clearing the site for rebuilding; the declining value of the dollar, etc. History shows us that property values, or more specifically building costs, rarely decline in this manner. Protection should therefore be arranged on the basis of resuming business as it was before the damage took place. This can normally be done only by insuring for the replacement cost of the property.

Replacement Cost Coverage

This type of coverage is more expensive than cash value coverage, since the insuror must set the premium sufficiently high not only to cover the estimated likelihood of a fire occurring, for example, but he must also try to anticipate the rate of increase in the cost of labor and materials in the reconstruction of the building in whole or in part.

Other Coverages

Even insuring at replacement cost will not, as a rule, cover the full cost involved in a major disaster. Business interruption, extra expenses, site clearance, intervening passage of new and more exacting building codes—all add to the already inflated costs of replacing the existing structure so that business may resume as before as rapidly as possible.

There are many endorsements available to extend coverage to fully compensate for all expenses involved in replacement. They should all be considered in the light of individual needs. Obviously, the greater the coverage, the higher the cost of coverage; but this cost must be weighed against the risk and its consequences.

INSURANCE PROBLEMS

The Small Business Administration has shown that 25 percent of businesses in this country have had some kind of problem with property insurance in a twelve-month period. These problems include cancelled policies; refusals to issue or renew insurance; prohibitive rates; and limiting coverage to well below the cash value of insured property.

In inner city locations or in certain types of business, policies, when issued, substantially limit the insuror's liability—frequently to the point where the policy is virtually useless as support protection.

Federal Crime Insurance Program

These actions by the insurance industry have created enough concern in the small business community to call for some kind of remedial action on the part of the federal government. Effective August 1, 1971, the Federal Crime Insurance Program came into being. This program, which requires the participation of individual states, provides for federally funded crime insurance at reasonable rates, based on the size and accepted risk of the insured property. Coverage is limited to a maximum of $15,000.

In order to qualify for protection under this program, however, a business must establish certain minimum protective devices and procedures. The businessman must, in short, recognize that he can get the supportive protection insurance offers provided he makes at least minimal efforts to protect himself.

The program prescribes locks, safes, alarm systems and other protective devices, and establishes the kind of protection that various kinds of business must provide for themselves in order to qualify for this insurance. For example, gun stores, wholesale liquor and fur stores, jewelry firms and drug stores must all have a central station alarm system. Service stations must have a local alarm system, and so on. Small loan and finance companies, theaters and bars—businesses rated as high risk—are also eligible for insurance under the program.

The program still has a long way to go before it covers firms in every state, but its appearance on the scene is encouraging. Not only does it provide for insurance coverage of premises otherwise difficult or impossible to insure adequately or reasonably, but it also focuses attention on the very real need for the insured to take positive steps to provide protection of the premises to prevent loss, and to use insurance to defray those losses that do occur only when those security measures fail.

In short, it takes insurance from the front line of crime prevention—where it clearly cannot perform—and puts it into the reserve or back-up position where it can.

Chapter 6

OSHA

The atmosphere of reform at the turn of the century resulted in legislation which created, among other new laws, Workmen's Compensation. Even the most hard-headed employers found that their costs dictated compliance with the spirit of the law. As a result, it had a profound influence on the implementation of a long downward curve in work-connected accidents and injuries.

By 1958 this trend had leveled off. And by 1968, for the first time in over fifty years, the curve began to rise again.

14,000 occupational fatalities and over two million disabling work-connected injuries each year seemed to be considerably more than the number that might one day be arrived at as the irreducible minimum. The result, after much study, debate and controversy, was the passage of Public Law 91-596—The Occupational Safety and Health Act (OSHA) of 1970, which was signed into law on December 29, 1970.

REQUIREMENTS OF OSHA

General Health and Safety Standards

The OSHAct was not the first legislation to establish standards of occupational safety and health. The Walsh-Healey Public Contracts Act, the Federal Construction Safety Act, the Maritime Safety Acts, and the McNamara-O'Hara Service Contracts Acts all dealt with safety and health standards in specific fields and under specific circumstances. OSHA, however, is the first to attempt to legislate standards that will apply to virtually every employer and employee in the country.

General rules governing safety and health standards are laid down to apply to industry as a whole, supplemented by a vast literature relating to employee safety in virtually every industry or process.

These specific applications concern themselves with everything from the operation of machines and equipment to the handling of compressed air; from dry grinding and buffing operations to non-ionizing radiation protection.

Role of Records and Reports

Record keeping and reporting is an essential element of the Act, both for its enforcement and for developing information regarding the causes and the prevention of occupational accidents and illnesses. This latter function, although it may appear burdensome to some employers, is vital to OSHA's role in helping to accumulate data which will substantially reduce the annual toll of injuries and illnesses on the job, which has stubbornly refused to abate since 1958.

The great mass of material specifying the standards; the various reports to be made and filed; and the procedures that must be followed in posting notices, notifying employees, filing for variances, etc., may seem overwhelming—even impossible to comply with. Yet these requirements must be met. And a closer examination suggests that the routine is not as difficult as it appears. What it does require is planning to set up a reasonable system for compliance with the letter and the spirit of this important Act.

Extent of Coverage

In order to assure the broadest possible application of the Act, Congress worded it to state that every employer "affecting commerce" would be subject to its provisions. This means that if any devices, tools, materials, or equipment used on the job were manufactured in another state, the employer using such elements is "affecting commerce."

Thus almost all employers in every industry are covered. About 60 million workers in 5 million establishments are protected by the Act. And it follows that the 60 million workers so protected must comply with all rules, regulations, and orders issued pursuant to the Act that are in any way applicable to their actions and conduct.

Administration/Research and Development

Compliance with this wide-ranging legislation will be administered by ten regional and fifty area offices, under the policy guidance and supervision of the Department of Labor and the Department of Health, Education and Welfare.

In addition to enforcing the law, the National Institute for Occupational Safety and Health (NIOSH) under HEW is responsible for an ongoing program to develop criteria for establishing safety and health standards; to research health matters; to undertake special studies, make determinations and report on toxicity; and to undertake and promulgate professional training and educational programs.

In the pursuit of this research and development program, NIOSH has wide authority to require special reports and to interrogate employers and employees alike. However, this will probably be done only in the case of special studies, which are designed to develop data on specific industries or processes.

Standards for Toxic Agents

The OSHAct requires that the Department of Labor establish standards with respect to toxic agents or other harmful physical agents. These standards must be of such a stringency that, as far as is humanly possible, no employee will suffer any kind of impairment or disability no matter what his exposure (provided, of course, that all other safety precautions are complied with).

Variances

In view of the fact that standards are still being established in industries and processes where none have existed up to now, employers may, under certain circumstances, ask for variances from the standards eventually promulgated. These are:

1) *Permanent Variance*—The employer must prove that all the conditions, operations and processes which he is using will be as safe and healthful as those under the standard.

2) *Temporary Variance*—Interim variances from a standard for as long as two years may be allowed if the employer can prove that, although he is doing everything possible to protect his employees from the dangers covered in the standard, he is unable to effect immediate compliance, either because qualified personnel or materials are not available or because he cannot complete the necessary physical changes in time but is making every effort toward speedy compliance.

Notice to the employees that a temporary variance is being requested must be posted so that they may reply if so inclined.

3) *Experimental Variance*—In cases where an employer wishes to experiment to validate new techniques for safeguarding employees, a variance may be granted upon approval of the experiment contemplated.

4) *National Defense Variance*—Certain variances may be granted in order to facilitate or avoid impeding the national defense effort.

Predictions are that variances will not be widely granted—and then only after the most thorough examination of the existing conditions and an evaluation of the proposed operation.

Data Collection

In the past, certain information concerning illness and injury was submitted for insurance purposes under the provisions of Workmen's Compensation. Unfortunately, this material was neither complete nor compiled in a uniform manner on a national scale. As a result, its accuracy was suspect. It was neither useful nor used to develop an approach to illness and injury among workers.

Under OSHA the centralized collection of this information and its careful analysis is a major effort. The data will be used to identify the problem areas which up to now have been imperfectly catalogued. It will identify the types of problems, direct the countermeasures to be taken, and maintain an annual accounting of results.

Annual Records and Reports

In the early stages since enactment of the legislation, only a sample of approximately 250,000 employers are being asked to submit reports each year. These are firms falling within a statistical sample determined by the Bureau of Labor Statistics (BLS) and will be notified by the Bureau. Eventually virtually every employer of 100 employees or more will be required to submit these reports routinely.

It should be noted that, under the various laws governing health and safety requirements for workers on government contracts, contractors had responsibilities for record keeping. All of these requirements are now under OSHA exclusively.

Accident and Injury Reports

Under the Act there are other records that must be kept by *all* employers. These records report on all work-related deaths, injuries and illnesses. The records must be kept up to date, and they must be available to Labor Department Compliance Officers on request. They will normally be maintained on the forms provided, but they may be recorded on some private equivalent form providing the data is as readable and comprehensible as the form itself.

1) *OSHA Form 100*—This is a continuous log of all recordable injuries and illnesses. The log must be maintained in each work establishment, and each injury or illness must be entered in it as early as possible, but in no case later than six working days after being notified that such a case has occurred.

2) *OSHA Form 101*—This is a supplementary record for more detailed information on individual accidents. This record, too, must be completed and available for inspection no more than six working days after receiving information of a recordable case.

3) *OSHA Form 102*—This is an annual summary of injuries and illnesses and must be complete and ready for any examination no more than one month after the close of the calendar year. This form must be posted in a conspicuous place (where notices to employees are normally posted) no later than February 1 and must remain posted for 30 consecutive calendar days.

    These records must be kept in the work establishment for five years following the end of the year they relate to.

4) *OSHA Form 103*—This is the survey form used by BLS for the furtherance of its statistical studies. As noted, this information is required to be submitted by only a selected sample of employers at the present time. Eventually all employers of 100 or more employees may be required to submit this information. Should any employer receive Form 103 from BLS, he must complete it and return it to the Bureau.

It should be pointed out by way of clarification that Forms 100, 101 and 102 are for records only. They are not submitted to OSHA. They must be accurately recorded and kept up to date at all times, however. They must be available at the work establishment for inspection. Failure to do so will result in substantial penalties.

Other Requirements

In addition to complying with specified standards and keeping records, every employer covered by the Act must:

- Keep a careful record of all employee exposure to potentially toxic materials or harmful physical agents that are required to be monitored. Employees must be able to observe both the monitoring and the records. In the event that any employee has been exposed to these materials or agents in concentrations beyond those established by the standards, he will be notified immediately.

- Post a notice informing employees of their protection and obligations under the Act. Copies of this poster are available on request from OSHA regional and area offices. He must also have available to all employees a description of the standards relative to the particular establishment. These standards should be available in the production manager's office to any employee who wishes to inspect them.

- Post any citation issued for a violation. This citation must be prominently displayed at the place where the violation occurred.

- Notify employees of the application for a variance from the standards. This must be posted on the bulletin board and a copy delivered to authorized employee representatives.

False statements by an employer or by employees may result in a $10,000 fine or 6 months imprisonment, or both. In addition, any failure to post notices as required will result in civil penalties for the employer.

Inspection

In order for OSHA to be effective, there must be some inspection for violations of standards of health and safety, and a means whereby these violations will be mandatorily corrected and the law enforced. This is particularly important where the legislation is so wide-ranging and, as in this case, where an entirely new concept of standards has been established.

Violations stemming from ignorance, misunderstanding, oversight, or willful non-compliance can be expected, especially when we consider that these regulations apply to millions of workplaces that have never before been obliged to meet any—or at best the barest minimum—safety standards.

The provisions of the law regarding inspection for the purpose of identifying violations and enforcing compliance are tough. They allow few opportunities for compromise. The law is designed to eliminate, insofar as possible, all possible dangers to health or safety that exist under present standards, and to collect data to revise, update or tighten those standards to reduce such dangers even further in future years.

Rules Governing Inspections

In order to accomplish the goals of the Act, Department of Labor Compliance Officers are authorized to enter any premise at reasonable times *unannounced.* They may examine machinery, processes, materials, equipment, and the overall conditon of the work establishment. They may question any employer, employee, contractor, or sub-contractor, either in public or in private at the sole discretion of the inspector. Interference with the inspector in the performance of any examination he feels necessary is subject to stiff penalties.

It is significant to note that these inspections are emphatically designed to view the operation of the facility in its day-to-day condition. Every effort has been made to prevent employers from correcting violations or performing temporary cosmetic routines for the benefit of a pre-announced visit by an inspector. Except in certain rare instances, any advance notice of an inspection is subject to a fine or imprisonment, or both. If for any reason an employer is given authorized advance notice of an inspection, he is—

under pain of penalty—required to notify the employees of the fact.

Priorities for Inspections

Certain inspections can be anticipated (even though they will not be announced) according to priorities for their scheduling. These priorities are:

1) Catastrophes and incidents resulting in any fatalities or five or more hospital cases.

2) Employee complaints.

3) High injury and health hazard industries.

4) Routine random inspections.

Employee Complaints

Whenever an employee or group of employees file a formal, written complaint specifying circumstances that would indicate a violation, there must be an investigation. If none is made, or if no citation is issued after an investigation, the government is required by law to respond and to explain its position fully in writing. It can be anticipated that this may well become the principal motivation for future inspections.

High Hazard Industries

Companies with an accident or illness rate considerably higher than its industry norm can expect inspections. So can most, or perhaps all, of those companies that deal with lead, carbon monoxide, asbestos or similar health-hazardous materials. Statistics may begin to identify other materials as hazardous from time to time; companies dealing with such materials will then come under Labor's examination.

Nature of the Inspection

The inspection itself is characterized by extreme openness at all times. Before it begins, the inspector and the employer will meet to establish the inspector's identity and to discuss the purpose and scope of the inspection. At this time the inspector will show copies of any complaints his office has received. (Any employee complaint must be submitted to the employer at the time of the inspection;

however, the complainant's name may be blocked out at his request.)

Both employer and employee representatives may accompany the inspector on his rounds. In the event that employees have filed a complaint, they may employ the services of technicians or experts (at their own expense) to represent them during the inspection. Employees may designate how many of their group will walk through the facility, subject to the inspector's agreement that the designated employees are necessary to a thorough and open examination. (He will usually keep the party down to a minimum.) Employee time off from regular duties to accompany the inspection party is strictly an intramural problem. The government has made no provisions requiring that the employer pay or not pay for the time so spent.

The employer should describe the safety activities of the company to the inspector, and he should specify the voluntary efforts that have been undertaken to conform to the standards. Whatever else he does, the employer must note everything that occurs during the inspection in order to analyze the validity of the inspector's views and to determine whether or not to dispute citations and penalties.

Employer Cooperation

No employer may refuse entrance to the inspector unless he has a good—and probably extraordinary—cause for the refusal. If he is denied entrance or interfered with in any way, the inspector will obtain an inspection warrant and conduct his examination under the additional authority of that instrument.

Inspectors may not be required to sign any kind of release or waiver to enter the workplace. This includes any forms which the employer may normally require to be executed with respect to trade secret or security areas. The inspector may and should enter all such areas in the course of his tour. Management and employee representatives accompanying him to these areas may be restricted to those persons normally having clearance.

Inspections, except in special instances requiring the approval of the area director, will not be made in plants involved in labor disputes such as strikes and work stoppages.

Cost of Non-Compliance

These inspections can be relatively routine or traumatic for the employer, depending on how he has prepared his facility to comply with the OSHAct. He may have decided that the odds for an inspection were very small—after all, the fiscal 1974 budget called for 80,000 investigations of complaints in the nation's 5,000,000 covered establishments, or only 1.6 percent. He may have been unaware of some of the problems in his plant, or he may have failed to fully understand the provisions of the Act. In any case, such an employer could find himself in great difficulty if he is inspected. Violations can carry penalties of up to $10,000 or six months in jail, or both.

Violations

Violations of specified safety standards must be promptly cited in writing by the inspecting officer. He must at the same time fix a reasonable period for abatement or correction of the condition creating the violation. Employers and employees may appeal the length of time allowed for the abatement by application to the area director.

Violations are classified as follows:

1) *Serious Violation*—In any situation where there is a reasonable possibility that serious physical harm or death could result from the conditions or practices in operation, a serious violation exists. Such a citation provides for a mandatory fine of up to $1,000 for each violation. An employer can be excused from the penalty (though not the obligation to correct the condition) if he is able to establish that he could not have known of the condition after a conscientious effort on his part.

2) *Non-Serious Violation*—Cases where violation of a standard would not cause serious injury but could have an effect on the health or safety of employees are adjudged non-serious violations. These will be evaluated in terms of the number of employees exposed to the hazard, or the seriousness or nature of the injury or illness that might result. The number of incidents of violation would also be taken into account.

3)   *Willful Violation*—If an employer consciously and intentionally engages in an operation or practice which he knows to be a violation, he may be cited for a willful violation. It is important to note that the commission of the violation need not be established as malicious or hostile. It need only be proved that it was voluntary and intentional, as opposed to accidental or even negligent. It is possible that a single offense might not be judged "willful." Repeated offenses in these circumstances almost certainly would be.

Willful violations may be penalized by a fine of not more than $10,000 for each violation. If death to an employee results from such a violation, there could be a six-month jail sentence in addition to the fine.

4)   If several instances of a violation of the same standard exist, they will be regarded as one violation. The number of such instances will, however, have an effect on the penalty.

5)   Generally speaking, violations will be cited only when they are actually observed by the inspector. This will not apply in cases where it can be established that a violation existed and contributed to an accident.

6)   There is no provision in the Act for the issuance of citations or the levying of penalties on employees who fail to comply with standards relevant to their own conduct. Even though compliance to rules, standards and orders is required of each employee, wherever there is non-compliance it is the employer who is held responsible and is subject to citation. If, for example, an employee refuses to wear protective glasses, the employer may be cited. It is essential, therefore, that employers make compliance with all safety and health regulations and standards under the OSHAct a condition of employment.

7)   Circumstances frequently exist where more than one employer and his employees are engaged in the same or a related job or area. Responsibility for violations

in such a situation is frequently very difficult to assign. However, no employer is relieved at any time of the responsibility for the safety of his own employees.

If an employer creates or permits a violation, and that violation affects either his own or other employees in the same area, he will be cited. On the other hand, if an employer knows, or should know, of an existing violation and still permits his employees to work, he can be cited, even though he had no direct responsibility for creating the violating condition.

This means that employer "A" could be cited for creating or permitting a violation, while employer "B" is similarly cited, not for any responsibility for the violation but because, in effect, he overlooked the danger it might create.

Of course, if two or more employers jointly create or permit a violation, both will be cited.

8)   *De Minimis Violation*—In situations where a violation would have no probable effect on health or safety, a notice of *de minimis* violation will be issued without penalty. This notice will be mailed to the employer for his information, but he is neither required to post it nor to report corrective action taken, since any such action in this case would be at his own discretion.

Except for the *de minimis* notice (which is not classified as a citation), all citations must be prominently posted at or as near as possible to the place where the violation occurred. They must remain posted for a period of three days or until the condition is corrected, whichever is longer. The employer may post such comments as he wishes to make alongside the posted citation.

Appeals

Citations or penalties may be appealed by applying to the Secretary of Labor within 15 working days of receipt of the notice. The cited employer must also, within this period, notify the area director that he wishes to appeal. If he does not appeal within this time period, the citation and penalty become final and cannot be reviewed by any court or agency.

Employees who feel the assigned abatement period is excessive or unreasonable may file with the Secretary within a 15-day period. The OSHA Review Commission will conduct a hearing into any such complaints.

Corrective Action

After a citation has been issued and become final, and the abatement period has been established, the employer must take corrective active. He has no alternatives. At this point there is no avenue of review or appeal. If he fails to abate within the prescribed period of time, he may be assessed a penalty of not more than $1,000 for each day that the violation continues beyond the time allowed.

Urgent Conditions

If an inspector should discover a condition of such gravity and immediacy that it could cause death or serious injury before it could be handled through routine procedures, he must immediately notify affected employees and the employer. If corrective action is not immediately taken, the Secretary of Labor may petition the U.S. District Court for a restraining order. This order may specify and require immediate remedial action. It may also prohibit employees or others, except those involved in correcting the condition, from entering the threatened area.

INJURIES AND ILLNESSES

Any illness, injury, or death must be entered on the appropriate records and, in some cases, reported. Since this record will be inspected, it is important that procedures be established for handling this recording function efficiently. In all cases the employer should establish for himself (a) that there indeed was an injury, (b) whether it was work-related, and (c) whether it should be recorded.

Classes of Injuries and Illnesses

The classes of injuries and illnesses can be broken down as follows:

Fatalities—Any death resulting from on-the-job or job-related accidents or illnesses must be reported within 48 hours to the Area Director. It may be reported by phone or telegram if that would expedite the matter. The date of death must be entered in the log and in the annual summary.

Multiple Injury—All accidents resulting in the hospitalization of five or more employees must be reported within 48 hours to the Area Director.

Lost Work Days—Any injury or work-connected illness which results in lost work days must be recorded. This computation is not confined to those work days when the employee is bedridden or recuperating at home. It also includes days when the employee is assigned to a temporary job because of his condition; days when, although assigned to his regular job, he cannot perform all its functions; and days on which he is unable to put in full time on his regular job.

It is the employer's responsibility to determine the status of the employee with regard to his ability to perform in whole or in part. In many cases, he may wish to be advised by a physician. Each case should be investigated to insure the accuracy of its recording in the correct category.

It is sometimes difficult to determine when days are no longer lost as a result of accident or illness. As a guide, some criteria may be helpful:

- If a worker is transferred to another job, even though he still could not perform the original, his lost days are terminated.

- If a worker continues in his old job with the work redefined, eliminating the need for those parts he can no longer perform, lost work days are no longer recorded.

- If the worker leaves the company entirely, he is no longer employed, hence no longer officially losing work days.

There are many grey areas in this recording procedure, and in many cases the determination of the best way to file the data will not be easy. When in doubt, consult the OSHA Area Director.

Medical Treatment—Any case requiring medical treatment, but not otherwise recorded, must be filed. Medical treatment means care provided or administered by a physician. It does not include cases "which do not ordinarily require medical care," such as minor scratches, cuts, burns, blisters, and so forth, even though a physician may have provided treatment.

Employees are not required to report injuries or illnesses within any given period of time. However, the employer must record the case (if he determines it is recordable) within six working days after he learns of it.

Non-Fatal Cases Without Lost Work Days—Any illness or injury must be reported if it results in termination of employment, loss of consciousness, any restrictions on work or motion, or transfer to another job.

Diagnosed Occupational Illness—Any illness diagnosed as occupational must be recorded.

Occupational illness is any abnormal condition, other than one resulting from injury, caused by any kind of exposure to those factors attendant to an occupation. These are normally chronic illnesses caused by contact, inhalation, absorption or ingestion. Categories of such illnesses are:

1) Occupational skin diseases or disorders.

2) Dust diseases of the lungs.

3) Respiratory conditions due to toxic agents.

4) Poisoning or systemic effects of toxic materials.

5) Disorders due to physical agents other than toxic materials. These could include heatstroke, frostbite, effects of ionizing radiation such as X-ray, effects of non-ionizing radiation such as welding flash or microwaves.

6) Disorders due to repeated trauma. These might be noise-induced hearing loss, or conditions caused by repeated motion, vibration or pressure.

7) All other occupational diseases, including anthrax, brucellosis (undulant fever), infectious hepatitis, food poisoning, etc.

Records Maintenance

Because it is difficult to analyze the accident and illness experience among the varied activities undertaken by many (especially larger) companies when records are kept company wide, and because it is difficult to analyze these rates by size of establishment when the records are handled by central offices, employers are directed to keep these records *at the lowest level of operations* at the work site or establishment where the workers are actually employed.

This has a practical necessity, since these records must be available at the work site for examination by the compliance officer.

Form 102—the annual summary—must be posted at the work establishment to provide employees and managers at the local level with a picture of their own achievements or problems in safety and health, rather than a more impersonal presentation of the position of the company as a whole.

DESCRIPTION OF STANDARDS

Generally speaking, OSHA requires that an employer provide a safe and healthful place of work for his employees. This is spelled out in great detail in the Act to avoid leaving the thrust of the legislation in any doubt.

It speaks of free and accessible means of egress, of aisles and working areas free of debris, of floors free from hazards. And it specifies fire protection by fixed or portable systems, clean lunch rooms, and adequate sanitation facilities. Whereas in past years employers might contend in all sincerity that their facilities met community standards for safety and cleanliness, with the enactment of OSHA these standards have been formalized to describe minimum levels of acceptability.

Examples of Basic Standards

A few general descriptions of the areas covered, setting forth basic standards for employers in general, provide a useful sampling of the specific provisions of the OSHAct.

General

1) Passageways, storerooms and service rooms must be kept clean, orderly and sanitary.

2) Floors must be maintained clean and dry. Where there are wet processes, drains and false floors or mats must be provided.

3) Working areas and floors must be kept free from protruding nails, splinters, holes, and loose boards.

4)   Safe clearances must be allowed in aisles and working areas where mechanical handling equipment is used.

5)   Permanent aisles and passageways must be marked.

6)   Open pits, tanks, vats, and ditches must be protected by covers or standard guardrails.

7)   Floor loads, approved by local authorities, must be prominently displayed. Maximum loads may never be exceeded.

Exits

Every building designed for human occupancy must be provided with exits sufficient to permit the prompt escape of occupants in case of fire or some other emergency. They shall be free and unobstructed and clearly visible, or the route to them must be clearly marked. They must be readily accessible, free to discharge directly into a street, yard, or other open space that gives onto safe access to a public way. They must also be illuminated and marked by a readily visible sign.

Scaffold Safety

The footing or anchorage for scaffolds shall be sound, rigid, and capable of carrying the maximum intended load without settling. Unstable objects such as barrels, boxes, loose bricks, etc., shall not be used to support scaffolds or planks.

Scaffolds and their components shall be capable of supporting at least four times the maximum intended load.

Sanitation—General

All places of employment must be kept clean and orderly and in a sanitary condition.

Workplace and personal service rooms must be constructed and maintained to prevent the entrance or harborage of rodents, insects, and vermin of any kind.

An adequate supply of potable water must be provided in all places of employment. Drinking water must be available within 200 feet of any regular work location. Portable dispensers must be refilled and cleared daily. Ice for drinking water must be free from contamination.

The common drinking cup is prohibited.

Toilet Facilities

Every place of employment must provide adequate toilet facilities. They must be readily accessible and within 200 feet of all regular work locations.

Water closets shall be provided according to the number of persons employed on the premises:

1	to	9 persons	1 water closet
10	to	24 persons	2 water closets
25	to	49 persons	3 water closets
50	to	74 persons	4 water closets
75	to	100 persons	5 water closets
	Over	100 persons	1 for each additional 30 persons

An adequate supply of toilet paper with holder must be provided for every water closet.

Urinals may be provided, reducing the number of water closets required on a one-for-one basis, provided that the water closets are not reduced beyond two-thirds of the minimum specified.

Washing Facilities

Adequate facilities for maintaining personal cleanliness must be provided in every place of employment. They must be convenient for the employees and maintained in a sanitary condition.

Lavatories, supplied with hot and cold water, must be provided on the same ratio as water closets.

Whenever workers must wear special protective clothing because of exposure to excessive dirt, fumes, vapor or moisture, change room equipment with storage facilities separate for each employee and separate for street and protective clothing must be provided.

Lunchrooms

Where employees are permitted to lunch on the premises, the area used must be sanitary, although it may be in the work area. Such rooms shall not include toilet rooms or toxic material areas. No food or beverages may be stored in toilet rooms or toxic material areas.

Safety Code for Marking Physical Hazards

1) *Red* will be the basic color of
   - Fire apparatus
   - Danger signs
   - Emergency stop bar, buttons or switches.

2) *Orange* will be used to designate dangerous parts of machines or energized equipment.

3) *Yellow* will be the basic color for designating caution and marking hazards.

4) *Green* will designate "safety" and first aid equipment other than fire-fighting first aid equipment.

5) *Blue* is the basic color for caution, limited to warning against starting, using, or moving equipment under repair.

6) *Purple* will designate radiation hazards such as X-ray, alpha, beta, gamma, neutron, proton, dentron, and meson. Yellow should be used in combination for markers such as tags, labels, signs and floor markers.

7) *Black, white* or combinations must be the basic designations for traffic and housekeeping markings.

Fire Protection

The selection of extinguishers will depend on the nature of the fires anticipated, the construction and occupancy of the property, the nature of the hazard, and ambient temperature conditions.

The number of extinguishers needed is determined by the area and the arrangement of the building, the severity of the hazard, the anticipated classes of fires, and the distances to be traveled to reach each hazard.

SUMMARY

The above is merely a sampling of the coverage of the Act. It represents only that portion relative to those standards that apply generally to all workplaces. Specific requirements for machines and equipment, standards for materials, and power source standards are detailed and extensive.

Underlying all of them is the implicit view that society will not now accept the needless and careless risks which have, in the past, been considered unavoidable in business and industry.

The impact of the passage of the OSHAct has been tremendous. There can be no question that it will be even more startling over the next few years. In many ways it represents a social revolution that cuts across every element of society, affecting all workers whether they are covered immediately or not.

Yet the basic message of OSHA is not really very unusual at all. In its long and careful way, it says that every worker has the right to work in a safe and sanitary environment; that he has a right to be protected from injuries and illness caused by correctible hazardous conditions; and that he is obligated to cooperate in his own protection.

This is the message that every employer must hear if he has any interest in avoiding fines, the threat of jail, and endless suits and litigation.

Supervising Safety

Every manager should make every effort to strengthen the organization administering his health and safety program. Since it can be assumed that every employer by now has surveyed his operation to eliminate areas of non-compliance, it should be emphasized that he must continue to press for the highest safety standards by active administration of the safety aspects of the plant and regular re-surveys of the entire operation.

He must encourage his supervisors to check and re-check those areas where violations could occur. He must institute a broad comprehensive program of health and safety education. Both supervisors and employees (who are encouraged to an unprecedented level of participation by the Act) must be trained in safety and health regulations and techniques.

The better the understanding of everyone's rights and obligations, as well as the methods to exercise them, the better the program will work. This is an area in which all management is concerned from one viewpoint or another. From design to accounting, from R-and-D to merchandising, no part of the company is untouched by OSHA.

Since it impinges on so much of company policy, the responsible or coordinating element in an OSHA program should be a key member of management. He must have upper-echelon authority, since he will frequently have to adjudicate between departments. The safety professionals are essential to the program, but they will no longer administer it. That supervision must come from the highest levels of management.

Value of Form 100

In analyzing jobs, materials, processes and procedures to pinpoint problem areas, managers will find the OSHA log (Form 100) beneficial. The log can point out operations which need better control or reveal trends or patterns that should be encouraged or corrected. It can indicate where changes or modifications should be made. It can reveal the strengths and weaknesses in supervision, and help to identify those supervisors who are performing well or badly. It may suggest areas where more education or training is required. And it may serve as an indicator of employee morale.

A Change in Viewpoint

The threat of accident or illness has never improved any company's operation. Most employers have lived with a certain level of risk because they didn't think anything could or should be done about it. That is no longer possible.

If the traditional question—"Do you want it quick or do you want it safe?"—does not disappear entirely, it will have a different dimension—and require a different answer. What yesterday might have been thought of as an eminently acceptable risk may today be a very dubious one.

From now on, employers, grudgingly or willingly, will be testing the theory that a business will improve substantially when accidents and the threat of them are reduced or eliminated. Improved morale leads to increased production and work efficiency. Technical improvements frequently result from the obligatory overhaul of a production line or process. Certainly there should be a saving in insurance costs, and a substantial reduction of costly and damaging lawsuits.

Today, every employer is in the safety business up to his neck. If he doesn't swim, he will surely sink like a stone.

PART III
CAREER OPPORTUNITIES
IN SECURITY

Chapter 7

THE SECURITY
FUNCTION IN BUSINESS
AND INDUSTRY

The review of security techniques and systems undertaken in the preceding chapters has a broad application to business and industry in general. Security problems become apparent in virtually every area of a given company's activities. The need to deal forcefully and systematically with these problems has become increasingly evident to the industrial and commercial community, and steps have been taken by greater and greater numbers of these organizations to create a security effort as an organic element of the corporate structure.

THE EMERGING SECURITY FUNCTION

As career opportunities in this expanding field grow, so do the requirements for men and women with expanded managerial skills as well as a broad base in the theories and practices of security administration. Where and how this newly emergent security manager operates within the organizational framework, and how this relates to the total security system of individual concerns, depends upon the needs of that organization. General principles will apply throughout much of the business community, but specific applications must be tailored to the problems faced by each enterprise.

The Security Manager's Role

Directing our attention to the generalization of the security operation and the manager's role in it, we can find many common elements that are significant. In its organizational functions security encompasses four basic activities with varying degrees of emphasis. These are:

Managerial—Includes those classic management functions common to managers of all departments within any organization.

Among these are planning, organizing, employing, leading, supervising, and innovating.

Administrative—Involves budget and fiscal supervision, office administration, establishment of policies governing security matters and development of systems and procedures, developing of training programs for security personnel and security education of all other employees, providing communication and liaison between departments in security related matters.

Preventive—Includes supervision of guards, patrols, fire and safety personnel, inspections of restricted areas, regular audits of performance, appearance, understanding and competence of security personnel, control of traffic, condition of all security equipment such as alarms, lights, fences, doors, windows, locks, barriers, safes and communication equipment.

Investigative—Involves security clearances, investigation of all losses or violations of company regulations, inspections, audits, liaison with public police and fire agencies, classified documents.

It is important to remember that these latter three functions must be carried out to further the organizational needs of security. It follows that, in order to perform effectively, the security manager must be thoroughly conversant with all of the techniques and technologies inherent in such functions. But, in order to achieve the stated goals or the projected ends of the organization, he must be sufficiently skilled in his managerial duties to effectively plan, guide and control the performance of his department.

He cannot remain, as has been true so often in the past, merely a "security expert"—a technician with a high enough degree of empirical or pragmatic information to qualify him to undertake certain basic preventive or investigative tasks. The more he involves himself personally in such jobs, the more he will neglect his managerial functions. Security's role in the operation will suffer accordingly.

It is important to remember that companies which recognize the need and the efficiency of incorporating security as an organic part of their enterprise have begun the process of creating a new organizational function that will, along with such traditional functions as marketing, production, finance and personnel, play a significant role in the daily as well as the projected destiny of the company.

In this light it is clear that the security manager will function as an indispensable member of the staff. His role will extend far beyond the time-honored one of principal-in-charge of burglar alarms and package inspections, to which he has so often been relegated. This is not to suggest that there is any trend toward development of a power base for security management, but rather that many enlightened, modern company managers have assigned a higher priority to integrated security systems in an effort to encourage the growth of this function as an essential element of the firm's survival.

The Integration of Security

Once management has recognized that existing problems—real or potential—make the introduction or enlargement of security a necessity for continued effective operation, it is obliged to exert every effort to create an atmosphere in which security can exert its full efforts to accomplish stated company objectives. Any equivocation by management at this point can only serve to weaken or to ultimately undermine the security effectiveness that might be obtained by a clearer statement of total support and directives resulting in intra-company cooperation with security efforts.

Obviously the management of the security function and its goals must be compatible with the aims of the organization. This means that security must provide for continued protection of the organization without significant interference with its essential activities. Security must preserve the atmosphere in which the company's activities are carried on by developing systems that will protect those activities in much their existing condition, rather than attempting to alter them to conform to certain abstract standards of security. When the overall objectives of any organization are bent and shaped to accommodate the efforts of any of the particular functions designed to help achieve those aims, the total corporate effort inevitably becomes distorted and suffers accordingly.

Authority of Security

The degree and nature of the authority vested in the security manager become matters of the greatest importance when such a function is fully integrated into the organization. Any evaluation of the scope of authority required by security to perform effectively must consider several factors, both formal and informal, that exist in the structure.

The immediate urgency for increased security must be considered along with the status, growth and prior performance of the security effort. The peculiarities of the company itself in the context of existing intra-company relationships, whether by design or natural evolution, must be a factor. The potential for growth of the company and the attendant growth of staff activities should also be considered.

Ultimately, management will have to determine the costs and the projected effectiveness of the security function. Then it will have to face the big question as to whether security can be truly and totally integrated into the organization at all. If, upon analysis, it is found that the existing structure would in some way suffer from the addition of new organizational functions, alternatives to integration must be sought.

These alternatives usually consist of the application and supervision of physical security measures. This inevitably results in the fragmentation of protective systems in other areas requiring security. These alternatives are sometimes effective, especially in those firms whose overall risk and vulnerability are low. But as the crime rate continues to climb, and as criminal methods of attack and the underground network of distribution continue to become more sophisticated, anything less than total integration will become increasingly more inadequate.

Relation to Other Departments

Every effort should be made to incorporate security into the organizational functions. It must be recognized, however, that by so doing management creates a new function that, like personnel and finance, among others, cuts across departmental lines and enters into every activity of the company.

Security considerations should, ideally, be as much a presence in every decision at every level as are cost decisions. This will not mean that security factors will always take precedence over matters of production or merchandising, for example, any more than specific price factors will always determine decisions in these same areas. But security should always be considered. If its recommendations are overridden from time to time—sometimes a wise decision, where the cost or disruption involved in overcoming certain risks is greater than the risk itself—this will be done with full knowledge of the risks involved.

Organizationally, the relationship between security and other departments *should* present no difficulty. The interface serves to solve potentially disruptive or damaging problems shared by both functions. The company's goals are achieved by the elimination of all such problems. In practice, however, this harmony is not always found. Resentment and a sense of loss of authority can interfere with the cooperative intra-departmental relationships that are so vital to a company's progress.

Such conflicts will be minimized where security's authority is clearly defined and understood.

Definition of Authority

It is management's responsibility to establish the level of authority at which security may operate in order to accomplish its mission. It must have authority to deal with the establishment of security systems. It must be able to conduct inspections of performance in many areas of the company. It must be in a position to evaluate performance and risk throughout the company.

All such authority relationships, of course, should be clearly established in order to facilitate the transmission of directives and the necessary response to them. It should be noted, however, that these relationships take many forms in any company, not infrequently including an assumption of a role by a member of the organization who becomes accepted as a designated executive simply by past compliance and by custom. In such cases, where management does not move to curtail or redefine his authority, he continues in such a posture indefinitely, whatever his formal status might be. It is management's responsibility to continually reassess the chains of authority in the interests of efficient operation.

"Functional Authority"

In general, the security manager can be considered to serve a staff function. Traditionally this means that, as the head of a specialized operation, he is responsible to a senior executive or (in the fully integrated organization) to the president of the company. His role is that of an advisor. Theoretically, it is the president who, through his authority, implements the activities suggested by his advisors.

This is not always the practice. By the very nature of his expertise, the security manager has authority delegated by the senior executive to whom he reports. In effect, he is granted a part of the authority of his line superior. This is known as "functional authority."

Such authority rarely appears on a table of organization, since it is delegated and can be modified or withdrawn by the superior. In the case of security, this functional authority may consist of advising operating personnel on security matters, or it may and should develop into more complete functional responsibility to formulate policies and issue directives prescribing procedures to be followed in any area affecting the security of the company.

Most department managers cooperate with security directives readily, since they lack the specialized knowledge required of upper echelon security personnel and they are generally unfamiliar with the requirements of effective supervision of security systems and procedures. It is nonetheless important that the security manager operate with the utmost tact and diplomacy in matters which may have an effect, however small, on the conduct of personnel or procedures in other departments. Every effort should be made to consult with the executive in charge of any such affected department before issuing instructions that implement security procedures.

Levels of Authority

Obviously, there are many mixtures of authority levels at which the security manager operates.

His functional authority may encompass a relatively limited area, prescribed by broad outlines of basic company policy.

In matters of investigation, he may be limited to a staff function in which he may advise and recommend or even assist in conducting the investigation, but he would not have direct control or command over the routines of employees.

It is customary for the security manager to exercise line authority over preventive activities of the company. In this situation he commands the guards who in turn command the employees in all matters over which he has jurisdiction.

The security manager will, of course, have full line authority over the conduct of his own department, within which he, too, will have staff personnel as well as those to whom he has delegated functional authority.

SECURITY ADMINISTRATION

In his managerial function, the security director is responsible for planning. This means that he must establish objectives, analyze risks, allocate resources within prescribed or authorized budgetary limitations, determine what is to be done, how it should be done and how soon it should be set into operation. His plans should also include the formulation of broad general policies, and these should be committed to writing.

Organization of Security

Establishing Controls

The manager is responsible for organizing the security effort. This includes the establishment of controls over procedures such as shipping, receiving and warehousing, inventory, cash handling, auditing, accounting, etc.

Since all of these functions are performed in other departments, the most effective and efficient method of implementing such controls is by the presentation of a control or accountability system to the department manager and allowing him to express his views and to make counter-suggestions. There is no reason to suppose that a totally satisfactory control procedure cannot be reached in this spirit of mutuality. Only when such controls break down or prove to be inadequate should the security manager or his deputy step in to handle the matter directly.

As discussed in earlier chapters, such controls would also cover all physical protection devices, including interior and exterior barriers of all kinds, alarm and surveillance systems, and communication systems.

Identification and traffic patterns are other necessary controls. Identification implies the recognition of authorized vs. unauthorized personnel, and traffic in this context includes all movement of personnel, visitors, vehicles, goods and materials.

Organizing the Security Department

The security manager will be obliged to organize the structure of his department, including a careful description of duties, responsibility, authority, and the hierarchy of command.

In this area he will be responsible for the procurement of security personnel. This must be preceded by a careful analysis of personnel needs to implement plans previously drawn. Job descriptions must be developed and labor markets must be explored. Whatever specifications are arrived at, it is important that security personnel above all must be emotionally mature and stable people who can, in addition to their other skills and training, relate to other people under many conditions, including those of stress. It is also important to look for those persons whose potential is such that they may be expected to advance into the managerial ranks.

Supervision

In addition to planning, establishing controls, organizing a department, and hiring personnel, the fifth area of the security manager's responsibilities is supervision of security. And it is in the handling of this function that the entire security program will prove to be effective or inadequate.

The security manager must maintain close supervision over communications within his own departmental structure. It is essential that he communicate downward in expressing departmental directives and policies; and it is equally important that he receive regular communication up the organizational ladder from his subordinates. He must regularly study and analyze the channels of communication to be certain that the input he receives is accurate, relevant, timely, concise and informative.

Additionally, the security manager must set up a system of supervision of all departmental personnel to establish means of reviewing performance and instituting corrective action when it is necessary.

He must, above all, lead. His qualities of leadership will, in and of themselves, prove ultimately to be the most effective supervisory approach.

The Image of Security

The sixth element with which the security manager must deal is the image or representation of the security function.

In order for security to be effective in any organization, it must have the implied approval and confidence of that organization. Every time a guard acts the stern father figure, and every time a system is installed which is cumbersome and inefficient, the image of security suffers.

It is the task of the security manager to undertake a regular program of indoctrination to clearly define the role of security and of security personnel within the organization. Since employee participation and cooperation are essential to the success of any security effort, it is of extreme importance that a thorough indoctrination program eliminate any tendency to alienate these important allies by overbearing or bullying attitudes on the part of security personnel. Such a program must also impress upon all security people the importance of their role in public relations, both as employees of the company and as members of the security department.

No matter how well the department is organized, it cannot be effective without the full support of the people in the organization it serves. To achieve this support, the department must be educated in attitudes, duties and demeanor—and proper supervision must ensure that these attitudes are maintained. The very fact that security personnel are controlling the movement and conduct of other members of their community suggests that they must themselves be carefully controlled to avoid giving rise to feelings of resentment and hostility. The entire organization can suffer great harm as a result of general animosity directed at only one member of the security department who has conducted himself improperly or unwisely.

Departmental Evaluation

Regular departmental evaluations should try to determine whether security policies and procedures are being properly followed, and whether such existing policies and procedures are still desirable in their present form or should be modified to better achieve predetermined goals. These evaluations should also review all manpower and equipment needs and the efficiency of their current use in the conduct of the security program.

Since security concerns itself with prevention of damage, disruption or loss, its effectiveness is never easy to evaluate. The absence of events is not in itself revealing, unless there are accurate accounts of actual crimes against the organization during some prior, analogous period to provide a standard of measurement.

Even such a comparison is of dubious value as the basis for a thorough-going, objective analysis of security effectiveness. Circumstances change, personnel terminate, motivating elements alter or disappear. And there is always the haunting uneasiness created by the possibility that security procedures might be so ineffective that crimes against the company have gone virtually undetected—in which case the reduction or absence of detected crime presents a totally misleading picture of the company's position.

Cost-Effectiveness of Security

It is unlikely that any evaluation will ever determine the cost-effectiveness of any security operation. A low rate of crime—whether compared to past experience, to like concerns, or to neighboring businesses—is an indication that the security department is performing well. But how much is being protected that would otherwise be damaged, stolen or destroyed can be any figure from the total exposure of the entire organization to some more refined estimate based on the incidence of criminal attack locally or nationally, the average losses suffered by the industry in general, or the reduction in losses over a given period in the organization.

An estimate based on such figures might well serve as a practical guide to the usefulness of the security function. On the other hand, if a security operation costing $400,000 annually were estimated, by some formula using a mix of the data referred to, to have saved a potential in theft and vandalism of $300,000, would it be deemed advisable to reduce the department's operating budget by $100,000 or more? Obviously not. This would be roughly analogous to reducing or canceling insurance because damage or loss and subsequent insurance recovery for a specific period or incident were less than the cost of the premium.

Cost-effectiveness studies must be made of departmental operations even though they cannot be used, as a general rule, in devising a formula for computing the cost-per-$1,000 actually saved in cash or goods that would otherwise have been lost.

Management will, however, be responsible for regularly reviewing the savings in manpower that could result from the substitution of functionally equivalent electronic or other gear and the feasibility of taking such a step.

Personnel Review

In reviewing departmental performance, all records of individual personnel performance should be examined. The degree of familiarity of each man with his duties, and the extent of his authority, departmental and organizational goals should be examined. Corrective indoctrination should be required as indicated. The health, appearance, and general morale of each man should be noted.

Equipment Review

The state of all security equipment should be reviewed regularly with an eye to its current condition and the possible need for replacement, repair or substitution. This review should cover all space assigned for use by security personnel on or off duty; uniforms; arms, if any; communication and surveillance equipment; vehicles; keys; report forms, etc.

Carelessness or inadequate maintenance of such gear should be corrected immediately.

Procedures Review

A review of security department procedures is essential to the continued efficacy of the department. All personnel should be examined periodically for their compliance with directives governing their area of responsibility. Familiarity with departmental policies should be evaluated and corrective action taken where necessary. At the same time, the very usefulness of prescribed procedures should be reviewed, and changes should be initiated where they are deemed advisable.

APPLICATIONS IN BUSINESS AND INDUSTRY

In spite of the obvious advantages to integrating the security function into the organization as an organic function, many firms continue to locate this vital operation as a reporting activity of some totally unrelated department. Since, in many cases, the operation grew out of some security need that arose in a particular area, there is a tendency for that area to assume administrative control over it—and to maintain that control long after security has begun to extend its operational interests beyond departmental lines into various activities of the company.

In this way, security was traditionally attached to the financial function of the organization, since financial control was usually the most urgent need in a company otherwise unprepared to provide internal security. The disadvantages of such an arrangement are severe enough to endanger the effectiveness of security's efforts.

Functional authority cannot be delegated beyond the authority of the delegant. What this means is that, when the security manager gets his authority from the comptroller, it cannot extend past the comptroller's area of responsibility. If the comptroller can extend the role of security by dispensation of the chief executive, the line of authority becomes clouded and cumbersome.

Most business experts agree that functional authority should not be used to direct the activities of anyone more than one level down from the delegate in order to preserve the integrity of line functions Clearly, the assignment of security under the financial officer is a clumsy arrangement representing bad management practice.

Reduced Losses and Net Profit

Although security is a staff function, it could be viewed as a line operation—and one day may be. An effective security program, intelligently managed, can, by reducing losses, maximize profits just as surely as can a merchandising or a production function.

With crime rates soaring, currently costing all business approximately $16 billion annually, or about 1.5% of the Gross National Product, every business is targeted for losses. And all these losses come off the net profit.

Many managers, particularly those in retail establishments, push as hard as possible for an increase in gross sales. They frequently brush aside any words of caution about inventory shrinkage due to internal and external theft. New records in gross sales are their goals. Unfortunately, companies doing a gross annual business of from $50- to $75-million dollars have filed for bankruptcy.

It is the net that keeps business and industry alive. The gross may be a splashy figure. It may provide some excitement for the proud manager. But the net is the bottom line—and anything that eats away at that slender lifeline seriously endangers the organization.

Retailers, who are the hardest hit by criminal attack, may operate at a net profit of 2%. It has been estimated that the national average loss due to crime is approximately 1%—or one-half of.the retailer's profits gone into the pockets of thieves. A store netting 2% and doing a gross weekly business of $50,000 will net a little over $50,000 a year. But if that store is hit at the estimated national average, the owner will end up with $25,000 instead of the $50,000 really earned.

An effective security operation could cut those losses by as much as 90%. The savings between the investment in security and the additional earnings realized from reduced losses are net profit.

In this light, security can be seen as a vital function in the profit picture of any company. Any operation that can minimize losses and maximize net profits should clearly take its place as an independent organizational function, reporting to the highest level of executive authority.

Career Opportunities

The foregoing chapters have examined those basic principles of security that are generally applicable to business and industry. In the following chapters we will look at certain specific businesses— and, in the case of computer security, a specific area which is of increasing importance to almost any business of any size—as representative of the application of security systems, procedures and techniques to particular problems.

It would be impossible, within the parameters of an introductory view of security, to study all of the possible applications in all businesses. Industrial security has been a reference point throughout the earlier chapters. We have chosen, then, to give more detailed study to retail security, one of the prime targets of criminal attack; hospital security, representative of the protection and loss prevention function in a public service facility or institution; cargo security, which affects virtually all businesses across the board, and consumers as well, and which provides occasion for examining the special problems of protecting goods when they are not under control in a specific location; and computer security, which, as noted, is of ever-growing importance to the entire business and industrial community, as well as to government. Computer security could logically have been considered as part of internal security in that earlier chapter; it has been reserved for later in order to provide a more searching, in-depth discussion.

Finally, we will review security services, which are of rising importance in the ever-developing security field, and provide an alternative area of career opportunity for the student or professional in security.

From these discussions there should emerge a picture of the security manager who must necessarily be both problem- and systems-oriented. He will be emotionally mature, flexible, and open-minded. He has the capacity to bring out the best in his personnel; he can deal effectively with his superiors and his subordinates; he is a leader; he is aware of his public relations responsibilities, and he projects an image of competence and helpfulness. He has an unshakable integrity. He has the ability to understand and empathize with people, and he projects that openness. Quite capable of being wrong, he cheerfully admits his mistakes when they occur. He is aware, finally, that his is an important function in the achievement of organizational goals.

All this is really too good to be true; few people are all of these things. The security manager is, after all, only human. But, if he is an effective manager, he will possess many of these qualities. And, as can be imagined, a man with a significant number of these qualities is hard to find.

There is a very real shortage of manpower qualified to move into the management positions that need to be filled today in the security business. As more and more companies recognize the need for well-rounded people to work into the executive echelons, the shortage becomes ever more acute.

There is still much that needs to be learned about security—how it functions, and how it is best managed—but enormous strides have been taken in the last decade. The watchman or custodian of a few short years ago has given way to the energetic young man with a rounded education and a developed theoretical grasp of the nature and the function of security in an organizational context. The opportunities for a career in security management today are limited only by the energy and firmness of purpose of the aspirant.

Chapter 8

RETAIL
SECURITY

Most businesses are subject to problems of inventory shortages, but few feel the problem as acutely as the retailer. He necessarily deals in merchandise which must be received, stored, moved from warehouse or storage rooms to display areas on the selling floor. All of these operations pose a risk of loss from breakage or other damage, pilferage or quantity theft. Even inadequate or careless record keeping can effectively "lose" merchandise which fails to show up on proper inventory records. Merchandise on display is fair game for shoplifters during the day, and when the day is over the whole cycle begins again.

The three principal sources of loss to the retailer are the external losses from theft, the internal losses from employee dishonesty, and the losses that come from carelessness or mismanagement. Every area of loss must be counteracted in some way or the retailer may find that, while his gross business is booming, he is barely able to break even.

The arithmetic of it is as simple as it is familiar, but it bears repeating. If a supermarket operating at a net of 1% suffers inventory shortages from all sources amounting to $250 a week, it would have to have annual sales of $1.25 million worth of merchandise just to break even. Estimates of the Small Business Administration put the total loss to the retailing business from crime at $4.8 billion in 1971.

Generally, shortage control procedures apply to retailers of every kind—from all-night restaurants to department stores. Legal considerations, surveillance techniques, cash register control and the many other factors involved in a security program for a retailing facility are much the same. The details vary, and every establishment must ultimately make its own determination of what is best for its own application, but the basics are much the same throughout the trade.

Shoplifting

Retailing today demands that merchandise be prominently displayed and exposed to enable customers to see it, touch it, pick it up and examine it. There is little likelihood that merchandise will ever go behind the counter again in the great majority of stores that display it so invitingly. In this mode it is hardly theft-proof, but its theft must be controlled if profit margins are to be maintained. Shortages from shoplifting are not likely to be eliminated, but they must and can be reduced by a thoughtful and energetic security program.

Extent of Shoplifting

It is difficult to accurately pinpoint the full dimension of the problem presented by shoplifting. Few shoplifters are apprehended, and of those who are, even fewer are referred to the police. In fact, "the most optimistic studies of retail effectiveness in spotting shoplifters indicate that stores who do a good security job apprehend no more than one out of every 35 shoplifters."[7] Stores with vigilant personnel will report a relatively high level of shoplifting, while analogous concerns (in terms of store type, location, size, etc.) with indifferent anti-shoplifting programs report few such incidents.

A "West Coast study of 9,000 food store shoplifting cases indicated that more apprehensions are made in January than during any other month. Department stores, on the other hand, apprehend more shoplifters during the busy weeks before Christmas as well as during the week after the holiday. Unlike general larceny arrests, which according to national statistics peak in the month of August, a high incidence of shoplifting arrests is not reported during that time. However, most stores allow detectives to take vacations in August; those that retain a full detective staff during that month have found shoplifting apprehensions equal or exceed December arrests."[8] This experience has been frequently reported. It would appear that, where the vulnerability remains constant, the potential of shoplifting incident remains constant, ultimately varying only with the effectiveness of security measures.

An interesting analysis based on the experience of 632 super-markets in 18 Western supermarket chains suggests that shoplifting occurs at least six times per day per supermarket. The analysis further suggests that, since other reports have concluded that such incidence is considerably higher, by assuming only ten thefts per store per day and extending the experience to all chain super-markets, the national loss from shoplifting alone may be estimated at $450,000,000 annually.[9] This is considerably higher than the estimate of total shoplifting losses in all business in 1967-68 of $508 million arrived at by the Small Business Administration, since retailing represents only about 25% of the total losses covered by that study. The precise dimensions of the problem can only be guessed at, but whatever figures are used as a starting point for such guesswork, the total loss to shoplifters is huge.

Interestingly enough, this analysis shows that, in spite of publicity to the contrary, adults comprise 60% of those appre-hended and 75% of the loss in dollar value as a percentage of recovery. As between men and women, the division is almost exactly equal.

The same article reveals that, based on the only random study of department store shoplifting ever released, one out of each fifteen customers entering a downtown department store will probably steal; each theft will average between $3.69 and $7.15; less than one per cent will be apprehended; and age and race were not significant factors. It is interesting to note that men were involved in such incidents 25% less than women, and there seemed to be a correlation between frequency of incident and value of goods taken. The store with the highest frequency of observed shoplifting suffered losses averaging $7.15 per theft, whereas that with the lowest incidence lost $3.69. This test was confined to a study of downtown general merchandise department stores in major metropolitan areas and cannot be taken as representative of retailing in general; but the information may serve as a guide to more extensive conclusions.[10]

Methods of Shoplifting

Shoplifting is conducted in every imaginable way; the ingenuity of such thieves is legendary. By and large the great majority of thefts are simple and direct, involving nothing more sophisticated than putting the stolen merchandise into a handbag or a pocket. But there are certain methods beyond the simple taking of items that are in general use and should be anticipated.

Bags and packages are frequently used to conceal stolen goods. Bags may be those taken from the store victimized. In some cases items may be bagged by a clerk and stapled after the items have been paid for. The thief removes the staple, stuffs in stolen articles, and restaples the bag with a pocket stapler. Many items are packaged in such a way that they provide concealment for other stolen merchandise within them. Oversized cereal boxes, utensils, and unsealed boxes of bakery products are a few of the store-provided containers used to carry out small items.

Bulky clothing provides a number of possibilities for conceal-ment. Small items can be held inside the clothing under the arms; larger ones can be carried between the legs under a skirt or overcoat.

Ticket switching is common in all stores where the price of merchandise is marked on the item in some way. Sometimes tags are exchanged with cheaper merchandise; sometimes prices are partially erased or altered. If prices are marked with erasable markers, they can be changed easily to indicate a lower price.

Who Shoplifts

The shoplifter comes in all sizes and ages, and is of either sex. Generally they are broken down by type into the professional, the amateur, the drug user, the kleptomaniac and the thrill seeker. Of these the amateur is by far the largest in number.

The Amateur

The amateur comes from every economic group and represents every level of education. His thefts are generally impulsive, although a significant number of them find some kind of econom-ic or, more often, emotional satisfaction in their action, and they become virtually undistinguishable from the professional. The rest of them have no particular pattern of theft and may only steal once or, at most, a handful of times. Individually they do not represent a severe threat to the retailer, but the cumulative effect of such thefts, however motivated, can be very damaging.

The frequent repeaters ultimately become more methodical in their thefts and soon become a real problem. Among this group are those compulsive thieves known as kleptomaniacs—people who are unable to overcome their desire to steal. Such driven souls are very rare and do not make up a significant number or dollar loss to the retailing community.

The Professional

The professional shoplifter is a very real danger. His methods are well-planned and practical. He appears in every way as an ordinary shopper, carefully fitting into the environment of the store he singles out. He selects merchandise with a high resale value and a ready market. He is well connected with fences and lawyers. He is in every sense himself a supplier in the sub-system of illegal merchandising. He is the first man in the chain of underworld retailing and his activities are damaging in a number of ways to the legal storekeeper. Not only does he create severe losses for the legitimate retailer, but he sets up a system whereby he is effectively in competition with his victim's own goods.

The Drug User

The drug user, trapped in his addiction, must find a regular source of funds to supply his needs. He turns to many sources for his insatiable demands, but shoplifting is often the easiest and most lucrative. Thefts of $500 a day may be required to supply his habit. Normally merchandise is stolen for fencing at between 10% and 20% of its retail value. In other cases it may be stolen and returned for refund. Either way, the store suffers substantial losses.

The Thrillseeker

Thrillseekers are, more often than not, teenagers who shoplift as a gesture of defiance or under peer pressure to do something daring. The dollar loss attributed to juvenile shoplifting is probably somewhere between 25% to 30% of the loss from adult theft. Although the Uniform Crime Reports indicate that 66% of those arrested for larceny in 1969 were under 21, evidence with respect to shoplifting indicates that the number of juvenile offenders was about one third less than the adults apprehended.[11]

Preventive Measures

Surveillance

The key to successful shoplifting control is surveillance. Impulse theft, which comprises 95% of shoplifting incidents, is motivated by availability, desire and opportunity. Availability is a basic fact of life in modern retailing; desire is not only a private matter, individual in character, but cultivated by the merchant aggressively selling his wares. Neither of these factors can be controlled to a significant degree, nor should they be in retailing. But opportunity can be controlled.

The shoplifter characteristically snatches his loot when he thinks he is alone—when he is not being observed. An attentive sales staff occasionally asking if they can help, or rearranging merchandise in the vicinity of a customer acting in any unusual manner, can discourage most amateurs. Supervisors moving about the floor can also be effective in making known to the potential shoplifter that he may be observed at any time. Obviously such store personnel are primarily concerned with serving customers; but they can be an effective deterrent to shoplifting as well if they are aware of the problem and alert to any signs that might indicate a problem.

Mirrors

Convex mirrors, which are in wide use throughout the country, may be useful to avoid collisions of people or shopping carts rounding corners, but they have a limited use in the detection of shoplifters and may even hurt the program by creating an atmosphere of unwarranted confidence. Such mirrors distort the reflected scene in such a way that it is virtually impossible to see the details of action—and in shoplifting it's the detail of merchandise concealment that is of prime importance. Without a clear, precise image of what was taken and how it was concealed or where, it would be foolhardy and possibly costly in legal fees to confront a customer as a shoplifter.

On the other hand, flat mirrors of a decorative design might be built into the decor of the store at strategic spots which might otherwise be difficult to observe or keep under surveillance. Such mirrors, presenting a clear, undistorted image, can be very useful in the security effort.

Signs

There is considerable controversy over the use of signs warning of the results of shoplifting as a deterrent. Many merchants take the view that such signs are an insult to the great majority of honest shoppers who may become incensed and take their business elsewhere. Other merchants subscribe to the theory that such signs have no effect on honest people, since they clearly are not those to whom the message is addressed, but will remind those with larceny in mind of the gravity of their offense and thus deter them. There does not seem to be a clear-cut resolution of these viewpoints, except to say that there is no evidence anywhere that the posting of such warnings has ever had any effect on the incidence of shoplifting.

Displays of Merchandise

Merchandise displays must be appealing to attract customers, but they can also be secure to prevent theft. Symmetry in certain kinds of items displayed can be important in enabling the clerk or floor personnel to tell at a glance if any of the items are missing. Thin, almost invisible wires that in no way detract from the display can secure small items to the display rack. Dummy items look exactly like the actual merchandise and should be used when possible. Fountain pens can be displayed in a closed case, with two or three different models on counter chains outside for handling and testing. There are thousands of items and countless ways to display them to catch the customer's attention. Each such display must accommodate some means to provide security for the goods it presents.

Check-out

Check-out clerks should check all merchandise for signs of switched or altered price tags.

Customers should never be permitted to carry out unwrapped or unbagged merchandise. All purchases should be wrapped or bagged, and if the additional precaution of stapling is taken, the receipt should be stapled to the package.

Checkers in supermarkets must check shopping carts for merchandise on the bottom shelf. They should further check all merchandise purchased for possible concealment of other goods. This includes paper bags holding purchases of produce. Other items may be concealed beneath the apples or potatoes they contain.

Refunds

Refunds should be issued on the return of merchandise with the original sales slip. Since these are frequently misplaced, if the customer insists on a cash refund, full particulars including the name and address of the customer should be entered on the appropriate refund form. The original of the form, which has been signed by the clerk making out the form, the customer and the authorizing supervisor, should be presented to the customer for cashing at a refund window or other location handling such matters. Cashing of refunds should never be permitted at cash registers on the floor, since this practice could invite embezzlement by the operator of the register. A copy of the refund authorization should be turned in at the end of the day's business by the clerk who handled the transaction. All such refund forms should be numbered and accounted for, including damaged ones.

All customers must cash their own refunds, to avoid the possibility of forged slips supposedly being cashed by store personnel as a "service" to a nonexistent customer.

The refund system should be further checked by periodic audits of all refund forms used and unused and by contact with the customer so refunded.

Letters should be sent out regularly to a certain percentage of customers who have received refunds, asking if the service was adequate and if their request for refund was promptly and courteously complied with. If such letters are returned as undeliverable or if the customer denies having received a refund, an investigation of refund procedures is indicated.

Detectives

Professional shoplifters will not be deterred by the normal means that would discourage the great bulk of amateurs from stealing. Only a well-trained security staff can apprehend them. "Most specialists agree that a store should be staffed with at least one floor detective for every $4 million it earns in gross sales."[12]

Such detectives must learn to blend in with the normal routine of the average customer in a given store at a given time of day. He must learn the different patterns of the secretary on her lunch hour, the bored matron who shops to kill time, the energetic early morning customer with a specific mission in mind. He must observe the difference in pace of customers in the ten a.m. crowd and the hurry-to-get-home five-thirty shoppers. After learning the techniques of anonymity, the detective must learn what to look for—how to spot a potential shoplifter.

Spotting the Thief

A list of some of the signs to look for was developed by a large Midwestern store and includes the following:

"Packages: A great many packages; empty or open paper bags; clumsy, crumpled, homemade, untidy, obviously-used-before, poorly-tied packages; unusual packages—freak boxes; knitting bags; hat boxes; zipper bags; newspapers; magazines; school books; folded tissue paper; briefcases; brown bags with no store name on them.

"Clothing: A coat or cape worn over the shoulder or arm; coat with slit pockets; ill-fitting, loose, bulging, unreasonable, and unseasonable clothing.

"Actions: Unusual actions of any kind; extreme nervousness; strained look; aimless walking up and down the aisles; leaving the store but returning in a few minutes; walking around holding merchandise; handling many articles in a short time; dropping articles on the floor; making rapid purchases; securing empty bags or boxes; entering elevators at the last moment or changing mind and letting elevator go; excessive inspection of packages; examining merchandise in nooks and corners; concealing merchandise behind purse or package; placing packages, coat or purse over merchandise; using stairways; loitering in vestibules.

"Eyes: Glancing without moving the head; looking beneath the hat brim; studying customers instead of merchandise; looking in mirrors; quickly glancing up from merchandise from time to time; glancing from left to right in cross aisles.

"Hands: Closing hands completely over merchandise; palming; removing ticket and concealing or destroying it; folding merchandise; holding identical pieces for comparison; working merchandise up sleeve and lowering arm in pocket; placing merchandise in pocket; stuffing hands in pocket; concealing ticket while trying on merchandise; trying on jewelry and leaving it on; crumpling merchandise (gloves and merchandise).

"At counters: Taking merchandise from counter but returning repeatedly; taking merchandise to another counter or to a mirror;

standing behind crowd and taking merchandise from counters; placing merchandise near exit counter; starting to examine merchandise, then leaving the counter and returning to it; holding merchandise below counter level; taking merchandise and turning back to counter; handling a lot of merchandise at different counters; standing alone at counter.

"In fitting rooms: Entering with merchandise but no sales person; using room before it has been cleared; removing hangers before entering; entering with packages; taking in two or more identical items; taking in items of various sizes or of obviously wrong sizes; gathering merchandise hastily, without examining it, and going into fitting room.

"In departments: Sending clerks away for more merchandise; standing too close to dress racks or cases; placing shopping bag on floor between racks; refusing a salesperson's help.

"Miscellaneous: Offering questionable refunds; acting in concert—separating and meeting; setting up lookouts, interchanging packages, following companion into fitting room independently."[13]

Obviously, many of these are the actions of a perfectly well-intentioned person, but they may indicate a shoplifter, especially if several such indications appear in the actions of one person. In such cases surveillance is essential.

Arrest

There are many problems involved in making an arrest of a shoplifter. Chief among them is the possibility of liability for false arrest or unauthorized detainment. Since the laws of the various states vary so widely in their latitude in matters of arrest, it is essential that legal advice be sought to guide company policy in such matters.

Basically, however, the law governing shoplifting requires that a suspect (a) has appropriated store property with (b) a provable intent to steal. Both of these elements must be present to establish the fact that larceny in fact occurred. The actual appropriation must be established by the presence of the property in his possession and, even more difficult, intent to steal it must also be established, usually by the manner in which it has been taken or concealed on the person of the suspect. Care must be taken, both in the name of justice and in the interests of avoiding a damaging lawsuit, to differentiate between the absent-minded customer and the conscious thief.

Rumor to the contrary, it is not necessary that a suspect leave the premises with the goods to establish theft. If he has the item in his possession and the concealment on his person is of a sufficiently clear and evident nature to establish intent, he may be detained at any point. It is generally advised that the suspect's departure from the store adds weight to the evidence, but it must be borne in mind that the mere act of leaving with store property does not in itself establish intent.

Store policies on procedures of arrest and detainment must be predicated on legal advice and they must then be followed to the letter. As a general rule, it should be noted that in cases of detainment where neither a confession of guilt nor a form releasing the store from any liability has been signed, it will be necessary to prosecute to avoid the very likely suit for false arrest that could ensue.

Prosecution

The argument over when and whom to prosecute shows no sign of abating; certainly there appears to be little agreement among security professionals. The skew is probably more toward a tough attitude than otherwise, but every shade of opinion has its adherents.

Every study undertaken on the subject has shown that the rate of recidivism is enormous. The F.B.I. has published figures indicating that 83.4% of those arrested for larceny in 1969 had been arrested before for the same crime. Such a figure suggests that neither arrest nor the fear of arrest is a very effective deterrent.

This notion was pursued further by a study with highly significant implications. "A study of 400,000 case records of retail offenders who were either prosecuted or released by 30 leading stores in one city showed that one in four shoplifters repeated, some as often as 12 times. The discouraging findings induced a member of the local retailer's protective association to conduct a seven-year study of 400 shoplifters: 200 were selected at random as control subjects to be prosecuted or released according to the usual standards; and 200 were released on the condition they would place themselves under the care of a psychologist or psychiatrist. Both groups were followed up over a period of seven years.

"The 200 control subjects followed the expected pattern, with one in four of this group repeating the offense and being arrested over and over again by member stores. But of the 200 who sought professional guidance, in seven years not a single one was picked up a second time by any of the association's member stores."[14]

The results of this study would seem to point a way to a solution which neither stern prosecution nor indiscriminate release has provided. It would, further, seem to bear out the contention of the eminent Dr. Karl Menninger, who said: "My answer is that we, the designated representatives of our society, should take over. It is *our* move and our move must be a constructive one, and an intelligent one, a purposeful one—not a primitive, retaliatory, offensive move. We, the agents of society, must move to end the game of tit-for-tat and blow-for-blow in which the offender has foolishly and futilely engaged himself and us. We are not driven, as he is, to wild and impulsive actions. With knowledge comes power, and with power there is no need for the frightened vengeance of the old penology. In its place should go quiet, dignified, therapeutic programs for the rehabilitation of the disorganized person. If possible, the interest and help of a professional person should protect society during the offender's treatment, and should guide his return to useful citizenship as soon as this can be effective."[15]

CHECKS

The boom in private checking accounts which has been a part of the American economic scene for the past 25 years has led us to the point where business analysts and economists predict the cashless economy within the foreseeable future. At that time, they suggest, even checks will become passe, and all transaction will be based on debits and credits handled through instant recording of the exchange of goods or services, by centralized computers. Such computers would register a credit to an employee's account, credits where appropriate to various government agencies, and debit the account of the employer in the amount specified on payroll information. All obligations of the employee from mortgage payments to the expenses of a vacation trip would be handled in the same manner—by a simple series of ledger entries in an interlocking central computer system. Personal bank checks in use today are a crude and primitive version of this predicted streamlined system.

With personal checking accounts approaching the 100 million mark, and with over half the adult population handling its cash through such bank checks, it would appear that the physical handling of cash has been delegated to the banks. That day has not, of course, arrived, but it seems close. Even today almost 90% of all business transactions are handled by check.

Nature of a Check

A check is nothing more than an authorization to the holder of funds or bank to debit the account of the authorizer and credit the named person or the bearer in the amount specified. Provided that sufficient funds are available in debited account, the exchange is made either in cash or by crediting the account of the payee. This system is really only a version of the most rudimentary kind of bookkeeping accomplished by the exchange of tons of paper in a never-ending chain of authorizations.

Checks can and have been written on almost anything and in every conceivable form. Their legality as a negotiable instrument is in no way dependent on the usual check form in common use today. As a practical matter, many banks refuse checks other than those written on the forms they provide, simply because their procedure for processing the enormous volume they are called upon to record does not allow for personal eccentricities. The day when a Robert Benchley, a popular humorist of the 30's and 40's, could write checks in the form of risque poems on wrapping paper, is gone. Not because such a check duly signed and clear in intent is not a legal instrument for transferring funds, but because few people would cash it.

Advantages

Checks are in such wide use today because they are safe and convenient. Since cash can almost never be identified if it is lost or stolen, it is almost inevitably gone forever. Checks, on the other hand, have no negotiable value except as they are drawn and signed by the payor. Forgery is, of course, a problem, but that is considerably less of a risk than cash itself, which requires no criminal expertise of any level for its disposal.

Checks are invaluable as receipts of payments made and for recording of an individual's accounts for tax and other purposes.

Checks can theoretically be used only by the designated payee; therefore, if a drawn check is lost or mutilated, payment can be stopped by notification to the bank and a new check can be drawn. Checks lost or stolen in the mail can be handled in the same way, whereas mailed cash always is at considerable risk of non-delivery because of theft, loss or destruction.

Checks are a ready source of cash at any time. Many people are cautious about carrying substantial amounts of cash for fear of loss or theft. In the event a situation requiring funds arises, a check can provide the money necessary.

From the retailer's point of view, checks are a boon, since customers will buy what appeals to them at the moment and handle the transaction with the stroke of a pen. People who buy only with cash tend to be considerably more restrained in their buying habits. They are limited by the amount of cash they are carrying, and the psychological restraints imposed by the actual doling out of cash are considerably more persuasive than those involved in drawing a check. The operators of gambling casinos discovered many years ago that most players are cautious when handling cash, which has a reality that is automatically translated into food and doctor bills, car payments and the like. Chips in such establishments become simply tokens or scoring devices in a game and normal reticence is soon discarded. The check casher is usually a bigger buyer who allows his impulses greater sway in a retail situation.

Disadvantages

It has been variously estimated that somewhere between 70% and 80% of all bank checks are cashed in retail stores. It has been further estimated that almost 30 billion checks are now written annually. With about three quarters of these checks handled initially by retailers, it can be seen that the scope of this traffic is tremendous. Retailers are standing in for banks as the major supplier of cash in the country. Banks are in no way relieved of their heavy burden of accounting or of their responsibility as the ultimate repository of the funds drawn on by these documents, but they are no longer the primary source of supply of funds to the consumer. They have, in effect, thousands of agents or branch offices who perform line functions for them.

Need for Cash on Hand

This means that retailers must keep huge supplies of cash on hand to service check-cashing customers. This not only raises the risks of internal and external theft, but it serves to increase the cost of goods, since cash that might otherwise be invested in additional merchandise or other income-producing ventures must be held in reserve to handle the demands of customers.

Some economists take the view that this situation has its bright side, in that it forces many enterprises who might otherwise over-commit their funds to maintain a position of sufficient liquidity to protect against reverses. Nonetheless, many major retailers feel that their options are limited. Whatever the state of the economy, they must keep large supplies of cash available to provide what is traditionally a banking service.

Bad Checks

Bad checks of various kinds add to the costs of doing business. Forgeries or fraudulent checks are a direct loss of both merchandise and cash; checks which are ultimately collectable but are returned by the bank for any number of reasons present a huge administrative headache in making the collection. Even though a majority of such checks are soon made good, often after a single letter or phone call, the costs in such follow-up and double handling, as well as the additional costs of a collection agency if such is required, may bring the costs of each such returned check to between $3 and $5. A charge for any returned check is usually added onto the eventual collection, but most retailers feel that they do not collect the full amount such returns actually cost them in direct and indirect losses.

Types of Checks

There are many kinds of checks representing different types of transactions that may be presented for cashing to a retailer. Since eventually the store must establish a check cashing policy which will involve the kinds of checks it is willing to cash, a brief list of check types is in order.

The *bank check* is that most commonly in use. It is normally a rectangular piece of tamper-resistant paper that will reveal any attempts at erasure or alteration. On it are lines for date, the amount (in both script and numerals), the name of the payee and the signature of the payor. Most checks are personalized by the imprint of the name and address of the account holder, and they carry the name and address of the bank in which the account is held. These checks also have a space in which the check can be numbered by the payee for his accounting convenience, or they may be prenumbered by the bank. They will usually, nowadays, also have a certain amount of computerized information (such as account number, etc.) for the bank's convenience in processing.

These checks are precisely as good as the credit or the account of the person drawing them, assuming that person is who he claims to be.

Blank checks are simply forms which can be filled out as above, but they contain no information such as bank name, payor name or account information. They must be filled out to designate the bank on which they are drawn as well as other pertinent information. They must be hand processed by the bank and are therefore unpopular or frequently unacceptable to many banks.

Travelers checks are checks drawn to a certain designated amount and purchased from various banks and other agencies. At the time of purchase each check is signed by the purchaser. When they are cashed, they must be signed again in the presence of the casher to allow him to compare the signatures. Since these checks are serialized, they may be recovered in the event of loss or destruction, provided a record of these numbers has been retained.

Payroll checks are simply bank checks drawn to an employee. Since they are issued on the account of a reputable firm, they are usually acceptable in retail establishments although they can be and frequently are forgeries.

A *third party check* is one issued by one person to another. The merchant may cash such checks provided the designated payee endorses it as having been paid.

Government checks, whether for social security, tax rebates, salary or any of the many uses to which they are put, are essentially cash, except that the cash is assigned to a specific named person. Because the solvency of the payor is never in question, these checks have always had a lively traffic among thieves and forgers. It is particularly important that the identity of the payee be established and that these checks be examined for any signs of tampering.

Returned Checks

Generally speaking, checks are returned because of insufficient funds, closed account or no account in the bank on which they were drawn. Other problems such as a damaged check, illegible signature or amount or improperly or incompletely drawn checks may cause the bank to return them for clarification, but this latter miscellaneous group represents a small minority of the returns.

"Insufficient funds" returns are in the majority. Usually they indicate that the issuer is a bad bookkeeper, but he may also have in mind a loan from the store. He will cover the check eventually, but in the meantime he has the cash to tide him over.

"No account" and "closed account" are almost invariably fraudulent checks. It is possible that such checks were issued in error (such as transferring a bank account before all outstanding checks have cleared) but it is unlikely.

Approvals

The two most important and relatively simple steps that can and must be taken in approving checks are a thorough examination of the check itself and a positive identification of the check casher. Neither of these techniques will screen out the accomplished forger or a skilled bad check artist; neither will they eliminate the problem of the checks issued by honest but careless people which are returned by reason of "insufficient funds"; but they will reduce or possibly virtually eliminate a great deal of the most persistent losses.

Retailers are not obliged to cash checks. They do so in their own interests as a service. If a check looks dubious or the casher appears in any way to be other than what he presents himself to be, the check should be refused.

Examining the Check

Since, however, he will cash most checks presented, he must know what to look for. He must examine the check to verify the date, the amount to which drawn both numerically and in script, the name and address of the customer, the name of the bank, the signature of the customer, and the endorsement if it is a third party check of any kind, including payroll or government checks.

Identifying the Customer

He must also assure himself that the customer is properly identified. Generally, any official document which describes the holder and bears his signature can be accepted as adequate identification. One that also includes a suitably laminated (or otherwise affixed) photograph is even better. Such documents can be fabricated by an artful forger, but since forging of this kind requires some equipment and skill, it is not in any wide or general use.

A driver's license, passport, national credit card, birth certificate, or motor vehicle registration are all acceptable identifications. Club or organization membership cards, Social Security cards, hunting licenses and employee passes are not valid as dependable pieces of identification, since they are easily obtained or easily duplicated.

The best identification is by an authorization system established by the store itself. Credit cards or check cashing cards issued by the store provide a running record of the customer's account and serve as nearly positive identification for check cashing purposes. The cost of establishing such a system can be well worth it in stores suffering from substantial losses from fraudulent or uncollectable checks.

The information used to identify a customer should in all cases be entered on the back of the check for future reference. If, for example, a vehicle registration is accepted, its number should be entered on the check along with such other information as may be relevant.

ID Equipment and Systems

Equipment to record any check casher exists to help the retailer.

Camera devices which photograph the customer simultaneously with his check can be used to record the transaction.

Instruments to record his thumbprint on the check without the use of ink also serve in this capacity.

These devices are as effective as the system established for their use. If they are used carefully in accordance with a designated procedure, they can provide a useful record for the storekeeper. They do, however, tend to discourage check passing by thieves by their very presence, though they will by no means eliminate it.

Cooperative Systems

Cooperative systems among a group of participating merchants seem to be the most effective way of controlling large scale check fraud. An article in *Successful Retail Security*[16] describes such a system set up in both Yakima and Spokane, Washington, which managed to reduce losses due to check cashing substantially.

This approach involved an alarm system which warns member firms when a team of check passers arrives in town or when there is a report of a theft of blank checks that might be used. When such information is received from a police agency or from one of the member stores, the organization notifies certain retailers on their list, these retailers in turn have other retailers they are to notify, and they pass the word on to other designated merchants. Through this information circuit, all participants can be notified of the threat in a very short time. During its first year in operation, no professionally written bad checks were passed in Yakima and six bad check artists who tried were taken into custody.

Checks involving insufficient funds were also substantially reduced. Although the organization was not set up as a collection agency, it has been effective in recovery on such checks. Each member store reports on returned checks; the organization distributes weekly a list of all such outstanding accounts to every member, where it is posted by every cash register. No check will be cashed for customers with outstanding returned checks.

In order to effect recovery on returned checks, the customer is contacted by the organization and asked to set a date for taking care of the account. 90% of the checks are taken care of after the first contact. Sometimes a second contact is necessary, and the Yakima experience shows that 75% of the remainder make restitution by this second contact. After that, legal action by the merchants may be necessary to collect the remaining 2% of returned checks. The result of these efforts was a report by a local bank that the incidence of NSF (not sufficient funds) checks was reduced by 50% after the second year of this operation.

The author's conclusion from his experience in this enterprise was that any area could reduce bad checks by 85% if 70% of the merchants in that area joined together in such an association. He suggests that such an organization be non-profit with a salaried executive, providing a valuable service at low service charges. In his experience, fees charged ranged from $7.50 per month for stores with few bad checks up to about $25 per month for an average size supermarket, whose incidence of bad checks is necessarily higher.[17]

Store Policy

Every store must set its own policy for handling checks and maintain it. Certainly the policy should be reviewed and adjusted as necessary, but it must be strictly adhered to as long as it is in force. Employee indoctrination and continuing education in this problem is essential to the success of any program of check control.

As a guide, though certainly not as a rigid rule, certain limitations should be considered as follows:

- No third party checks. Payment can be stopped on such checks and the retailer's recourse is only to the customer, not to the payor.

- No checks on out-of-town banks. Such checks are difficult to verify, and the time involved in clearance is such that, if it is fraudulent, the customer has long disappeared before the check is returned.

- No checks over a certain amount above purchase.

- No checks cashed other than government or payroll checks.

- No checks cashed without adequate pre-determined documents of identification.

- No checks cashed drawn on other than personal checks imprinted with the name and address of the customer.

If such rules are followed, the loss from bad checks should be small, provided all checks are themselves carefully examined by the cashiers.

It is generally agreed that a retailer should never suffer losses greater than .005% of the face value of checks cashed, although some do, in fact, have losses as high as 1%. Such a drain on the resources of any company is clearly intolerable. It must be corrected by the establishment of reasonable controls and a program of employee education.

BURGLARY

According to the FBI's Uniform Crime Reports, burglary represented approximately 40% of all crime in the United States in 1971, and the incidence of burglary had increased by 163% in the years between 1960 and 1971. Even by the very conservative estimates of the Small Business Administration, burglary losses to businesses in this category amounted to almost $1 billion in 1967-8. Less than 20% of the burglars of all occupancies are apprehended, and the rate of apprehension is dropping a few percentage points annually. So with burglary the most frequent crime, with an increasing rate of incidence and a decreasing rate of arrest, it is essential that every businessman—the retailer in particular—take most particular care to protect himself against this attack on his assets.

The Attack

A burglary attack on a retail establishment is similar to such attacks on other types of facility, except that the job of the burglar is simplified by the ease with which he can inspect the premises before he makes his move. He can familiarize himself with the physical layout, store routines, police patrols and internal inspections, if any, simply by appearing as a customer. If he wishes to engage in a more exhaustive survey, he might even pose as a building or fire inspector and make a minute examination of alarm installations, safe location, interior lock construction and every other detail of the store defenses that may interest him.

Assuming he has found a weak point, the burglar enters through a door or window, through the roof, or possibly from a neighboring occupancy. Most successful burglaries (and less than 10% of detectable attempts are unsuccessful) are made by forced entry. Curiously enough, the greatest number are made through the front door or main entrance.

A considerably smaller number of burglaries involve the stay-in who gathers his loot and then breaks *out,* and is gone before guards or police can respond to an alarm.

Merchandise

Most burglaries involve the theft of high value merchandise, although any goods will do in the absence of the "big ticket" items. Police reports repeatedly show the most astonishing variety of goods have been targeted by enterprising thieves. Anything from unassembled cardboard cartons to bags of flour are fair game. Obviously such merchandise would hardly be considered as high risk assets, but they cannot be overlooked as possible loot. Burglars come in all shapes and sizes, and what may seem to one as cumbersome and unprofitable may be remarkably appealing to another.

Cash

Cash is naturally the most sensitive asset and the most eagerly sought, but, since it is usually secured in some manner, it represents the greatest challenge to the burglar.

Stores keeping supplies of cash on hand are particularly susceptible, especially before payday in areas in the vicinity of a company or companies with substantial payrolls. If such stores customarily cash payroll checks as a service to these employees, a burglar can assume that adequate cash must be on hand in anticipation of the next day's demands. Particular care must be taken in such cases if there is no way to handle cash needs other than to store it on the premises overnight. Every other avenue should be explored, however, before deciding to keep substantial supplies of cash on hand overnight.

Since any cash will normally be held in a safe, it requires some degree of expertise to get at it. Boring, jimmying, blasting and even carrying away the entire safe are the methods most commonly used. Few burglars are sufficiently skilled to enter a safe by manipulation of the combination, so applied violence is the normal approach. Unfortunately, even in cases where an attack is unsuccessful, the damage to the container is likely to be severe enough to require its replacement.

This is equally true in all areas where entry is made or attempted. The high cost of repairs of damage to buildings and equipment can be almost as harmful as the loss in cash or merchandise. A survey of 4,800 company-owned and franchised stores in a nationwide chain of convenience stores indicated that 42% of the burglary loss was reported to be in building and equipment damage and theft.

Defense Against Burglary

In order to reduce the threat of burglary, the premises must be secured against entry. The principles governing this hardening of the facility are essentially the same as those governing the protection of buildings of all types.

Physical Defense

Doors must be of a heavy construction, hung in a frame sufficiently strong to avoid prying. Hinges must be either located within the doorway, or the screws or bolts must be set in such a way that they cannot be removed. If doors contain glass panels, they should be protected by grilles or screens, or they should be of burglar-resistant or polycarbonate materials.

Windows should be of these same materials, or they should be barred or screened. In many applications where windows serve only to admit light, they might well be replaced with glass brick.

Roofs should be protected by chain link fence or barbed wire if there is a possibility of gaining access to them from neighboring buildings, trees, poles, or by climbing drain pipes, etc. Skylights should be protected by bars, grilles or fencing materials.

All means of ingress of 64 square inches or larger must be screened, locked or grilled in some manner. These would include air vents, manholes, loading chutes and all other avenues through the outer walls.

Locks, adequate to the job, must be installed. And the installation must provide for adequate seating of the bolt.

Keys must be distributed sparingly and kept under tight control at all times. They should be inventoried regularly, including a physical inspection of all keys held by authorized holders.

Adequate lighting is as important to the protection of a retail outlet as it is to an industrial facility. In the case of supermarkets with large parking lots, lights should cover the area, for the safety and convenience of the customers as well as for the deterrent factor such lighting offers after closing hours.

Landscaping should be designed to avoid casting shadows or creating concealed approaches to the store.

In urban facilities, interior lighting inside the perimeter, especially in the area of entrances of any kind, will help to deter the would-be burglar as well as to make his presence visible to passing patrols.

Fences and walls may be advisable in some situations, although their effectiveness must be reviewed in the light of the ease with which they can be scaled from nearby elevations.

Alarms

Alarms of some kind can make the difference between a really effective program and one that is only half safe. The type of alarm providing the best results is a matter of some disagreement, but there is little disagreement over the effectiveness of some kind of system, however simple.

Many stores report satisfaction with local alarm systems. They feel that the sound of the signaling device scares off the burglar in time to prevent his looting the premises. Many police and security experts feel such alarms are ineffective, because the response to the signal is only by chance and because such a device serves to warn the intruder rather than aid in his capture. Since most managers are less interested in apprehending thieves than in preventing their theft, and since local systems are inexpensive and can be installed anywhere, they continue in wide use. Certainly they are preferable to no alarm system at all.

Whether the outer perimeter should be alarmed and whether space coverage alarms should be used is a matter peculiar to each facility. This must be carefully studied and determined by the manager. But it is always important to alarm the safe in some way.

Safes

The location of the safe within the premises will depend upon the layout and the location of the store. If it is readily visible from the street, many experts agree that the safe should be located in a prominent, well-lighted position where it can be seen easily by patrols or by passing city police. In some premises—especially where no surveillance by passing patrols can be expected—it is generally recommended that the safe be located in a well-secured and alarmed inner room which shares no walls with the exterior of the building. Floors and ceilings should also be reinforced. The safe should be further protected by a capacitance alarm. The classification of the safe and the complexity of its alarm protection will be dictated by the amount of cash it will be expected to hold. This should be computed on the maximum amount to be deposited in the safe and the frequency with which such maximums are stored therein.

Basic Burglary Protection

Obviously the amount of protection—and the investment involved in it—will depend on the results of a careful analysis of the risks involved in each facility. There is no single way—no magic solution—to the problem of burglary. Every system and each element of that system must be tailored to the individual premise.

The manager must evaluate the incidence of burglary in his neighborhood as well as the efficiency and response time of the police. He must consider the nature of the construction of his building and the type of traffic attracted to his area. He must consider the ease with which his merchandise can be carried off and by what probable routes. He must consider the reductions in insurance premiums provided by various security measures. He must consider the advice of city police in anti-burglary measures and their experience in analogous situations. In short, the retailer owes it to himself to learn as much as he can about coping with the problem of burglary and dealing with it as forcefully, economically and energetically as possible.

ROBBERY

Robbery, next to larceny of $50 and over, is the fastest growing crime in the country. From 1960 through 1971 the number of incidents of robbery reported in the FBI's Uniform Crime Reports rose a staggering 259.5%. During the same period this crime increased from 59.9 per 100,000 inhabitants to 187.1 per 100,000 inhabitants, an increase of 212.4%.

And retailers take the brunt of this attack. Since less than a third of the robbers are arrested and, as in the case of burglars, this percentage is declining annually, the retailer is obliged to take measures to protect himself from this most dangerous crime.

Nature of Robbery

An interesting profile of robbery, which might be projected into a larger, national view of the problem, can be found in the study of 4,800 company-owned and franchised stores of a national convenience store chain referred to earlier.

In this study we can see few differences in the experience with robbery due to geographical location, and the occurrence of such incidents is fairly constant throughout the week. The daily peak of robberies is 10:30 p.m. However, those stores remaining open all night suffered the majority of robberies between midnight and three a.m. 96% of the money was taken from cash registers, and 81% of the robbers were armed with handguns. 87% of the assailants were under 30, and of these 97% were male. Although weapons were used (or the threat of harm inherent in a robbery), violence was actually carried out in only 4% of the cases. 80% of the death cases in this study occurred in situations in which store personnel had done nothing to motivate the attack. And while 60% of the robberies were carried out by a lone robber, 75% of the death or injury occurrences took place in situations involving two or more robbers.

Profile of the Robber

If we can take this microcosm as representative of the problem faced by the retailer, we find that the robber is most apt to be a young man working alone, carrying a pistol and threatening employees with bodily harm unless they hand over the money from the cash register. He rarely carries out his threats—probably because the employees wisely comply with his demands—unless he is accompanied by a partner, in which case he is very likely to cause death or injury without provocation.

Robbery Targets

The robber is a criminal making a direct assault on a person responsible for cash or jewelry. His target may be a messenger or store employee taking cash for deposit in the bank or bringing in cash for the day's business. Robbers frequently enter a store a few minutes before closing and hold up the manager for the day's receipts. They may break into the store before opening and wait for the first employees, who will be forced to open the safe or the cash room. In recent cases, managers or members of their families have been kidnaped or threatened in order to coerce access to company cash.

Delivery trucks and warehouses can also be targets for robbers. These latter cases require several hold-up men working together and, although they are a very real threat and do occur, the majority of cases are still carried out by the lone bandit attacking the cash register.

Businesses most frequently attacked are supermarkets, drugstores, jewelry stores, liquor stores, gas stations and all-night restaurants or delicatessens.

Prevention

Probably the major cause of robbery is the accumulation of excessive amounts of cash. Such accumulations not only attract robbers who will have noticed the cash handled, but they are more damaging to business if a robbery does occur.

It is essential that only the cash needed to conduct the day's business be kept on hand, and most of that should be stored in a safe. Cash should never be allowed to build up in registers, and regular hourly checks should be conducted to audit the amounts each register has on hand.

Every store should make a careful, realistic study of actual cash needs under all predictable conditions, and the manager should then see to it that he has only that amount plus a small reserve on hand for the business of the day. Limits should also be set on the maximum amount permitted in each register, and cashiers should be instructed to keep cash only to that maximum. Overages should be turned in as often as necessary and as unobtrusively as possible.

In large volume stores, a three-way safe is an invaluable safety precaution. Such safes provide a locked section for storage of two or three hours' worth of money that may be called upon for check cashing or other cash outlay. The middle section has a time-lock which is not under the control of any of the store personnel and has a slot for the deposit of armored car deliveries or cash build-ups. Cash register trays are stored in the bottom compartment.

Movement of cash build-ups from registers to the safes should be accomplished one at a time, and every effort should be made to conceal the transfer. This is particularly important at closing time when registers are being emptied. The sight of large amounts of cash can prove irresistibly tempting to a potential robber.

Cashroom

If the store has a cashroom, it must be protected from assault. Basic considerations as to location and alarming were outlined in the discussion on anti-burglary measures, but an additional precaution to protect against robbery should also be noted. The room itself must be secure against unauthorized entry during business hours. If possible, a list naming those persons authorized to enter the cashroom and under what conditions should be drawn up. It should be made clear that there are to be no deviations under any circumstances from these authorizations.

The door to the room must be secure and securely locked. If fire regulations indicate the need for additional doors, they should be equipped with panic locks and alarmed. The entrance should be under the control of a designated person who, through a peephole or other viewing device, can check persons wishing to enter. In larger facilities, entrance can be controlled by an employee at a desk outside the room, although this arrangement should be studied carefully before implementation, since a robber could force such an employee to permit entrance unless her position is protected from the threat of attack.

Opening Routine

Since the potential for robbery is always present when opening or closing the store, it is important that a carefully prescribed routine be established to protect against this possibility.

At opening, the employee responsible for this routine should arrive accompanied by at least one other employee. The second stands well away from the entrance while the manager prepares to enter. The manager checks the burglar alarm, turns it off and enters the store. He checks the interior, including washrooms, offices, backrooms and other spaces that might offer concealment to robbers. After a specific, pre-established time, he reappears in the main entrance and signals his assistant in a pre-determined code.

This code, which should be changed periodically, should be given both in voice and in some hand signal or gesture such as adjusting his tie or smoothing his hair. One coded reply and gesture should indicate that all is clear; another should indicate trouble. It is important that this code be innocent and reasonable sounding to allay any suspicions on the part of the robbers if any are present.

If the assistant gets the danger sign, he should reply with something to the effect that he will be along as soon as he gets a paper or checks his car or gets a cup of coffee. He should then proceed slowly and deliberately to a predetermined phone and make a call to the police. Experienced security men all stress the value of a card with the number of the police station typed on one side and coins to make the call taped to the back. For lack of a dime, stores have been robbed.

Messengers

Since a basic principle in robbery protection is minimizing cash on hand, money will necessarily have to be moved off the premises to a bank from time to time. Ideally this should be done by an armored car service. If this cannot be done because of cost or non-availability of such service, efforts should be made to get a police escort for a store employee acting as messenger.

In any event, the messenger so assigned must be instructed to change his route; he should be sent out at different times of day and, if possible, on different days of the week. He should regularly change the carrier in which the money is stored. It might be anything from a brown paper bag to a tool box. He should stay on well-populated streets and he should move at a pre-determined speed. The bank should be notified of his departure and given a close estimate of his time of arrival.

Closing Routine

Shortly before closing time, the manager makes a check of all spaces, much as at opening. An assistant unobtrusively watches for any unusual activity on the part of departing customers. When all registers have been emptied and the money locked up, he stations himself in the parking lot and watches as the manager completes his closing routine. When the manager has checked out the alarm and locks up, the assistant is free to go. If the manager signals trouble, during any of this routine, the assistant will follow the same routine as at opening.

Other Routines

Since any entry into the store leads to a potential for robbery, it is important that the manager never try to handle it alone. There have been a number of instances where hold-up men have called the manager and reported damage or a faulty alarm ringing in order to lure him into opening the store. He is then in a position to be coerced into opening the safe or at least providing a bypass of the alarm system. If he is so notified, the manager should phone the police and/or the appropriate repair people and wait for them to arrive before getting out of his car. He should never try to handle the matter without some back-up present.

Employee Training

A full cooperative effort by all employees is essential to any robbery prevention program. A properly indoctrinated employee will know how to use a silent, "hands off" alarm if such an installation is deemed advisable. He will learn to question persons loitering in unauthorized areas. He will be alert to suspicious movements by customers and will know what to do and how to do it if he is aware that a hold-up is in progress in some other part of the store. He will remain cool, make mental notes of the robbers' appearance for later identification. He will cooperate with the robbers' demands to an acceptable minimum. He will not under any circumstance try to fight back or resist. If possible, he will hand over the lesser of two packs of money if that choice is open to him, but he will never do so if there is the slightest chance that the robbers will become aware of what he has done. He will remain alert to the possibility of robbery, and, if he should become involved in one, he will make careful observations for assistance in apprehending the criminals.

INTERNAL THEFT

Methods of Theft

The list of ingenious techniques employed by dishonest employees to steal from the stores that hire them is endless. Whatever systems are installed to control inventory shrinkage from internal theft, some clever employee finds his way around them.

Although the report of the Small Business Administration conservatively estimates that employee theft represents about 12% of the losses suffered by all small business, the 1972 study by the Department of Commerce previously referred to (*The Economic Impact of Crime Against Business*) states that: "While shoplifting appears to be the most serious problem for retail establishments, most observers believe that because of the reluctance of businessmen to admit the magnitude of their employee theft problem, that figure is seriously understated. Some believe that employee theft accounts for substantially more loss than shoplifting by custom-ers."[18]

It is important for every manager to familiarize himself with some of the ways employees steal in order to better understand the scope and nature of the problem. Familiarity alone will not serve to stamp out this problem, but it will aid in focusing on those areas of greatest danger and help to establish some counter-measures.

Cash Registers

Perhaps the most widely employed method of theft involves some kind of juggling with the cash register. Since cashiers and managers both have access to customer cash to be deposited in registers which record the transaction, there are literally thousands of opportunities to manipulate the accounting before individual sales or even the receipts of the day are posted. Cash never recorded as received is clearly much harder to locate or identify as missing than cash or merchandise which has been entered and later stolen.

Methods used with the register usually involve the regular theft of small sums by under-ringing the amount received. This involves ringing $9 for a $10 purchase, for example, and pocketing the difference at the end of the day before checking in. An even cruder method is to ring up "no sale" or "void" instead of the amount paid from time to time, and pocket the cash received at the sale.

From time to time employees are found who have an opportu-nity to remove the tape from their register shortly before the end of the day. They put in a fresh tape and ring up all sales accurately. When they check in, they pocket all the proceeds on the new tape, destroy it and hand in the prematurely removed old tape as their record of the day's receipts. They have effectively gone into business for themselves on company time.

Several flagrant cases of such private enterprises have involved managers buying their own register and, either alone or in collusion with another employee, setting up an additional check-out lane during a few hours of heavy traffic. Such cases are unusual but not unheard of.

The Giveaway

Checkers in supermarkets have frequently been discovered giving away large amounts of merchandise to fellow employees when they check out with purchases by ringing up only a small fraction of the value of the merchandise. In other cases they have similarly accommodated friends or family members.

Many cases are reported where various store clerks set themselves up as traders, exchanging stockings for another clerk's shoes, or blouses for costume jewelry, etc. In a somewhat similar approach, employees have been found who, after selling an item for its regular price, enter it as having been sold to an employee at the regular employee discount price. The seller then pockets the difference.

Price Changing

Every kind of store must be alert to price changing by employees and customers alike. Employees have an opportunity to alter price tags and buy the merchandise on their own or in collusion with an outside confederate.

Vendor Kickbacks

Collusion with vendor in receipting more merchandise than is delivered is common in every business, and cannot be overlooked in retail establishments particularly.

Refunds

In cases where refund controls are inadequate, it is a simple matter for employees to write up refund tickets and submit them for cash either in person or through a confederate.

Merchandise Theft

It is common for employees to transport items of merchandise to their cars in the course of the day. This is often done in several trips or perhaps by accumulating items in one package to be removed just after the store opens or before it closes. If package control procedures are in operation, such employees may purchase an inexpensive item or two and pass out the package with the legitimate sales slip for authorization.

Stocking

Stocking crews have an excellent opportunity to steal enormous amounts of merchandise during the off-hours when they are normally in operation.

Embezzlement

Embezzlement in retail establishments takes the same forms as those discussed in the earlier chapter on Internal Theft, and the same precautions must be exercised to prevent it.

Countermeasures

Screening and Supervision

The first and most important countermeasure that every store should establish is a firm employment policy involving a careful screening of every job applicant. These procedures, which have been discussed in an earlier chapter on Basics of Defense, may seem burdensome and even more costly than the more *pro forma* checking used by too many stores, but in the long run they will pay many times over.

The next basic is enlightened supervision. No system of controls can be effective if it is not adhered to; and unless it is supervised it may soon fall into disuse. Supervision to confirm that all established procedures are being followed as well as regular audits of their effectiveness are vital to the success of any retail security program.

Shopping Services

Auditing of the efficiency, effectiveness and honesty of sales people by shopping service investigation has long been used by retail firms as one of the most accurate methods of determining the conduct of their operation. Such service is not inexpensive but has established itself as sufficiently effective in the reduction of theft and the improvement in the performance of personnel to pay for itself many times over.

The tests conducted by such services or shoppers usually take note of the employee's appearance, manner, helpfulness and salesmanship, as well as checking for any signs of dishonesty.

Dishonesty testing consists of the creation of situations where the employee could easily steal or fail to ring up cash; the employee is then observed or audited to check on performance under the circumstances. The situations created are in no way construed as entrapment or enticement but represent re-creations of normal situations which could be anticipated in the regular course of business. Since shoppers making these tests are unknown to the employees, their reactions to the tests can be taken as indications of their performance in similar situations with any customer. A full report is submitted after each such test for the manager's reference and review.

Most stores contract for such services on a yearly basis at a set fee, with the understanding that inspections will be made with a certain prescribed frequency. In this way store management has some confidence that an ongoing audit made by objective outside investigators will help to uncover inadequate or dishonest performance by sales employees and cashiers.

Controls

As in other types of business, the receiving and shipping area of any retail operation is a particularly sensitive one. Since the life blood of the business flows across these areas many times in the course of a year, it is essential that there will be full accountability for all movement of merchandise, that a perpetual inventory be an integral part of the system, and that the area be restricted to those persons specifically authorized to be there.

All merchandise loaded or unloaded should be subject to periodic spot checks, including a complete unloading and recounting of merchandise already loaded from time to time. Cargo seals should be secured and inventoried regularly. All broken shipments should be investigated and secured. All loading and unloading procedures should be supervised.

Maintain a rest room and lounge area for drivers which is separate from the facilities used by stockroom, warehouse or dock personnel, and insist on this separation. Do not permit drivers to enter storage or merchandise handling areas.

The effectiveness of procedures should be audited from time to time by introducing errors into various operations. For example, the number of cases to be shipped might be invoiced incorrectly to see if the checker catches the error; truck seals might be logged and noted on invoices incorrectly to verify the alertness of personnel involved.

Trash
Trash removal has always been a problem because it provides an efficient means of removing merchandise from the premises

without detection. It is important to have a supervisor on hand when trash is loaded for removal, and it is especially important that trash collection for such removal be separated from loading or receiving dock areas.

Package Control

It is important that some kind of control procedure be established to inspect packages removed from the premises by employees. Retail outlets have a particularly difficult time with this problem, since employees have regular access to large amounts of merchandise—much of it small enough for easy portability. Receipts must accompany every purchase removed and all packages should be subject to inspection as desired by security personnel.

Employees

The state of employee morale is the key factor in any store security program. If employees are totally familiar with all store rules and policies; if they feel that they are appreciated as human beings as well as store employees; if they feel they are an important functioning part of the organization, they will respond with increased efficiency and the problem of dishonesty will diminish. It is essential that management bear in mind that, along with its desire to receive reports up the ladder, it must reciprocate by communicating back down the structure. Communication must be a two-way path.

Employees should be motivated to perform, not compelled to. This can be accomplished only by clear statements of policy, firm supervision insisting on compliance, and intelligent leadership. In this atmosphere morale should grow and the company will prosper.

Chapter 9

HOSPITAL
SECURITY

Hospital security is a particularly exacting field for the security professional because it encompasses virtually every aspect of security. Assuming a 500 bed facility, there must be a feeding operation which will serve between 1,000 and 1,500 meals a day; there will be more than 20,000 patients with an average stay of slightly over a week each; there will be from 500 to 1,000 prescriptions filled daily, and these prescriptions will be administered from numerous nursing stations where substantial stores of narcotics are kept. Linens, surgical gowns and many of the staff uniforms will be laundered in the hospital laundry and then distributed to as many as 30 or 35 linen lockers, where they are further distributed as needed. Nursing supplies, from bedpans to water carafes, must be distributed throughout the hospital to be used as necessary. With an average of four visitors per patient per day, there is a regular daily traffic volume of from 1,500 to 2,000 unaffiliated outsiders moving through the facility, and they are present largely in concentrated groups during visiting hours. Such a hospital would employ a staff of nearly 2,000 men and women. Add to this the 500 or so doctors who are affiliated with the hospital, the 400 to 800 volunteers, and perhaps 200 or 300 student nurses. All of this activity is contained within a complex which must be protected from unauthorized intrusion, and yet which must be in operation on a 24-hour basis—an operation which must have on hand substantial supplies of substances such as anesthetics or oxygen, which are highly flammable.

The list of a hospital's necessary activities is an extensive one. It clearly establishes one fact: a hospital is faced with the security problems of a hotel, an industrial complex, an office—and it has its own special security problems as well. As formidable as the hospital security problem is, there are still administrators who fail to understand the concept of hospital security. Many are too concerned with the problems of patient care to feel that security plays a role in hospital administration. This attitude is all too frequently supported by the apparent absence of a real problem. Since hospitals generally are well behind industry in inventory control techniques, they are often unaware of their enormous losses in equipment and supplies. Other problems, such as theft from patients, vandalism, accidents and employee pilferage, are frequently untabulated and unclassified, so that no log or summary exists to outline the dimensions of the problem.

This is not to suggest that hospital administrators are totally blind to the security problems that beset today's hospitals. They are not. Many of them have established outstanding systems providing an integrated program of multifaceted protection of their facility. Others, however, have not dealt forcefully with the problem. Like some retail establishments, they are aware of inventory shrinkage, but, since they are unable to identify the problem or its true extent, they consider it a necessary cost of doing business in modern society and add it into operating overhead—a cost which the patient must ultimately bear.

Because of the mission of the hospital, the providing of care and help to sick or troubled men and women, in some cases or in some facilities without recompense, many administrators and much of the public think of hospitals as essentially immune to the security difficulties so prevalent in other sectors of society. The image of dedicated men and women working selflessly to bring ease to troubled minds and bodies in an isolated atmosphere remote from other realities is an historical one which the media do nothing to dispel. Certainly there are huge numbers of dedicated people working to alleviate human misery in every hospital, and clearly the end product of a hospital's effort is human welfare—but it does not operate in a vacuum any more than any other business does.

Whatever its product, a hospital is business. Big business. Hospitals are one of the largest businesses in the country today—with purchases of materials, supplies and services amounting to well over $4 billion annually. That much money will always represent a prime target for thieves, and it must be protected.

The cold facts are that there are just as many thieves, neurotics, malcontents and generally malicious people on the hospital staff, among the patients, and among the crowd of visitors as there are in any department store, for example, and they must be considered as part of the security risk. In addition to the mass of people having free and authorized access to the hospital premises, there is always the threat of violence to personnel, especially nurses who come off duty late at night; vandals; surreptitious intruders; dishonest vendors; armed robbers and arsonists—to run through a melancholy list of some of the types that could plague hospitals as well as other businesses and institutions.

A HOSPITAL SECURITY PROGRAM

In many respects a hospital's security program will follow along the same broad outlines that guide the security efforts of every facility. Due consideration must be given to parking areas, proper perimeter controls, adequate lighting, entrances and their uses, guard usage, security control points, communication, regular inspection procedures, and employee and visitor identification and control. These considerations, along with fire and safety procedures and the institution of systems to protect against internal theft, are analogous to basic security needs generally applied throughout the business and industrial community; but many of them must be shaped to the specific needs of hospital operation.

Administratively, the security function must be based upon broad—sometimes difficult—policy decisions by top management. As implied above, implementing necessary security policies is difficult for many hospital administrators, since they are frequently convinced that a total security program is not compatible with a patient care facility—an attitude that originates in an underlying conviction that such a program interferes with the primary mission of the institution and ultimately is unnecessary in the facility.

Unfortunately, such an attitude can lead to enormous waste as a result of employee theft or carelessness, theft from and even assault on patients, as well as countless other attacks upon institutions and their occupants. It is essential that the hospital administration be shown the need for a security system that will assist in smooth, safe and efficient operation rather than hindering it by restrictive authoritarian policy. It must also know that the elimination of theft, either internal or external, will allow the hospital to survive or expand rather than be forced into unwanted economies that could serve to reduce the services the institution would otherwise provide.

The objective of hospital security is a broad-based, integrated program that deals with security problems as a whole rather than attempting to fragment the effort by compartmentalizing it into areas where fire and safety, theft, pilferage, crimes of violence, disaster reaction, traffic control and security administration are treated separately. A well-integrated, properly coordinated program will be infinitely more effective and more efficient—both in terms of performance and economy—than an effort which attempts to deal with these problems under several supervisory areas.

Security Personnel

It would be virtually impossible to provide even rudimentary security for a hospital complex without a security force. The diverse nature of the necessary security program and the conditions under which the program must function make it mandatory that there be trained personnel to inspect, patrol, assist and administer a long and varied list of duties, areas and responsibilities. They cannot be replaced by locks, alarms or barriers.

The flow of personnel and supplies and the admission of patients continues around the clock. There is never a time when the facility is locked up and secured. This traffic must be supervised and controlled—usually at more than one point. Such control can be effectively managed only by personnel—and by personnel specifically assigned to this function.

Patrols

This control starts with the exterior areas, including parking lots, which must be patrolled periodically during the day and even more frequently at night. Since most hospitals are located in urban areas where crime rates are more than 800% higher than in non-urban areas, such patrols will play an important role in reducing vandalism, surreptitious entry, and damage or theft of cars in the hospital parking area.

Since many hospital facilities consist of numerous buildings, including living quarters for staff employees within a single complex, it is important that the entire area be well-lighted and that landscaping be so designed that shrubs, hedges and other ornamental plantings be located well away from walk-ways (especially those that might be used at night), and that approaches to such possible concealment be particularly well-illuminated. In the case of those hospitals where the complex is broken up into two or more units and where hospital personnel must use public thoroughfares to travel from one unit to another, a more extensive patrol will be required.

Many hospitals today routinely provide security escort service for nurses going off duty late at night. The exact nature of this service depends, of course, on the location of the hospital; whether the nurses generally leave singly or in groups; whether they are housed in a nearby nurses' residence or are taking public transportation to their own homes. In some cases a regular patrol along a short route to the nursing residence will suffice; in others it will be necessary for a security officer to walk nurses singly or in groups to the nearest bus stop. In any event this is an important aspect of personnel protection in the security system of every large urban hospital.

Entrances

A few modern hospitals have been built in such a way that traffic control is considerably simplified by the design of the buildings themselves. Unfortunately this is very rare. Most of today's hospitals make no architectural concessions to security considerations. A high percentage of them have been expanded over the years by the addition of new wings or annexes. This has frequently complicated the exposure of the facility by creating new entrances, passages and receiving areas which can be difficult and costly to control.

It is of prime importance that a careful survey be made to determine how entrances can be reduced to an absolute minimum. The ideal arrangement, which would consist of a main entrance essentially for ambulatory patients, attending medical staff, and vistors, an emergency entrance for non-ambulatory patients, and an employees' entrance, can rarely be achieved. But efforts should be made in that direction. Greater efficiency and control might be achieved by allowing use of certain employee entrances only during specified periods of the day and preventing their use at other than authorized times. It might be possible to close off various entrances at night and thus reduce the security exposure during this potentially dangerous period.

Each hospital is unique in design; each requires a thorough examination to determine the optimum number of entrances to satisfy the needs of efficient and effective operation, as well as to provide the maximum in protection from intruders and from employee theft. All such entrances should be manned by security personnel when they are open and in use.

Perimeter Openings

This same evaluation should be made of the perimeter if the hospital grounds are protected by a fence or other barrier. These entrances may not require a guard station unless there is a need to direct visitors entering the main gate to various buildings in the complex, but they should be reduced to the absolute minimum necessary for operation. In most cases where the facility occupies sizable grounds within a single unit, the main gate alone will suffice to handle all entry, including ambulance and pedestrian traffic. Additional gates which are normally locked should be provided for fire equipment in the event of an emergency.

Windows and Other Openings

As in the case of other facilities, ducts, ventilation and sewer pipes of more than 96 square inches should be screened, as should windows or other openings leading into storage spaces or other areas where surreptitious entry might be made. It is important to survey the exterior of the buildings with a careful view to detecting those vulnerable points where entry is at all possible.

Security Posts

During the day a security post should be established inside the main entrance at some convenient but prominent position in the lobby. From this post a security officer can issue passes, check identification badges, watch for unusual behavior on the part of visitors; and he can additionally be on call to assist in any situation within his area of responsibility. If the main door is secured at night, this same function can be moved to the emergency receiving area or it can be dispensed with during low traffic periods.

The Command Post

Most large hospitals have a permanent area which is manned 24 hours a day. Its location will vary from facility to facility, but it will be most effective, as a rule, if it can be located in a position that overlooks the principal employee entrance and, if possible, the shipping and receiving dock and its associated storage areas.

This post is the security command post. It may be one of several in a block of security offices. Space availability is always a problem in today's growing hospitals, but a central command facility is essential, as is adequate office space for the security director and sufficient area for files and records.

The central command post, if it is located as described, can check employee packages, issue identification badges, visually supervise the entrance and exit of employees, supervise time cards, maintain a key locker, maintain a regular log of the events occurring on each shift, and act as a lost and found office.

It will, more importantly, act as the nerve center of the entire security operation. All patrols will report in on a regular basis by phone or radio. All paging of security personnel will originate from this center. All calls for assistance or reports of suspicious or unsafe or emergency conditions will be made to this post. All alarm systems will record here, whether they also record elsewhere or not. The effectiveness of the operation will be greatly improved if a closed circuit TV system is included as part of the protective hardware, in which case the monitors and controls of the system would logically be located in this central command location.

Interior Patrols

Security personnel patrolling the hospital interior should be reachable by a pagemaster system. These units are simply palm-sized portable receivers which emit a tone when one is activated from the central command post. The security officer responds by phone. These are preferable to transceivers, since there is more privacy in a phone call. (Security patrols outside the main building or in areas away from patients and visitors use transceivers, since they are not concerned with the problem of privacy and since phones for fast response are not available in many of the areas being patrolled.)

During these interior patrols, the security officer must be on the alert for any number of situations or conditions that could prove dangerous. He will watch for intruders, unauthorized visitors, unidentified employees, sneak thieves, disorderly conduct, vandalism and any signs indicating an attack on locked spaces. He will assure himself that all doors that should be locked are properly secured. He will check fire hazards such as accumulations of trash, improperly stored flammables, smoke or the odor of smoke, smoking in no smoking areas, blocked exits, improper stacking in sprinklered storage areas. He will check fire extinguisher tags for inspection dates. He will note any safety hazards such as slippery floors or stairways, or holes in floors or stairs, etc. He will watch for running water, machinery or lights not turned off, or anything of a like nature that is wasteful of hospital resources and ultimately potentially dangerous.

Visitor Control

The continual flow of traffic which is organic to the function and the operation of a hospital is one of the principal factors which makes hospital protection unique in the security profession. Visitors, volunteers, student nurses, consulting physicians, and the regular discharge and admittance of patients make for a continually changing sea of faces. The great majority of this traffic is as unfamiliar with the hospital as it is to the security personnel. Coupled with this is the atmosphere of urgency and emotion so frequently present.

Visitors are concerned about the condition of a close friend or a member of the family; patients themselves may be nervous and apprehensive—in any event, not completely themselves. Allowances must be made for behavior under these conditions. Every hospital security program must stress the importance of understanding and tolerance of certain behavior. Training programs should include discussions of this problem for regular re-indoctrination of the security staff.

Uniforms and Weapons

In this same regard, many hospitals have found that the overall security effort has benefited from a positive effort to deemphasize or to totally eliminate the authoritarian police concept of the security service. Some have chosen to discard the police-type uniforms and outfit security personnel in blazers with an identifying pocket patch. While this is a uniform and identifies the function of its wearer, it does not suggest "police."

Many hospitals disagree with this policy and prefer instead the traditional uniform for its psychological deterrent capabilities. They argue that the presence of a uniform instantly suggests to the patient and the intruder alike that the facility is secure—that the premises are protected.

There is, at the same time, and largely from similar considerations, disagreement on the question of arming security personnel. Some hospitals feel that it is desirable to issue sidearms to all security officers during their duty hours; others arm only gate personnel and exterior patrols; and some issue no arms at all. Few, if any, hospitals, however, fail to have some weapons available for issue in the event of an emergency that would require them.

A Total Prevention Program

Whatever the policy governing uniforms and sidearms, it could be said that there is general agreement among hospital security directors that their job is to create a total, integrated prevention program—a program in which the security department spearheads the effort and bears the major responsibility for its effectiveness, but also one which calls upon the entire hospital organization for active participation. It would be wrong to consider the security department as being the sole element in the security of the organization. Rather, through education and training, every employee must become, to some degree, a part of the system.

By the same token, every security officer must be made aware that he is part of the total health-care function of the hospital. This can only be accomplished through continuous training in which security personnel are exposed to every facet of the hospital operation. Such training should go well beyond security procedures into all areas of concern and interest in the hospital. It could be of great value if such training sessions were addressed by doctors, nurses, dieticians, pharmacists and others on the hospital staff who can acquaint the security department with the functions and problems in all areas of the hospital's operation.

Badges and Passes

As a control factor it is recommended that all personnel be instantly identifiable by some means. Specific techniques vary from place to place, but the most widely accepted practice is the requirement that all employees wear an identification badge carrying his name, department, employee number, his signature and, in most cases, his picture.

The nature of the information on the badge will vary with the policy of the hospital issuing it. In some cases the badge may designate only the employee's name and department, and he will be required to carry on his person an I.D. card with more complete information as to his status. Badges may be color-coded to indicate department or authorization of the wearer. For example, volunteer workers will generally be asked to wear a badge, but the badge should be color-coded to indicate their volunteer status. Contractor personnel who will be on the premises for any appreciable length of time should be issued badges to be prominently displayed, but they too should be color-coded to immediately indicate the wearer's position.

Attending medical staff is usually required to wear some identification even if it carries no more information than the name of the wearer. They should, however, be required to carry full identification on their person. Doctors are not always remarkably amenable to such procedures as badging, but a well-planned public relations program can go a long way toward convincing them of the necessity of such controls, which must include everyone to be effective.

For such a program to work, it must be enforced. Supervisors in every department must be convinced of its need, and their full cooperation must be sought in order to maintain the program's integrity. In addition, the security department should make spot checks from time to time to see that badges are being worn.

Visitor Passes

Many hospitals have found it advisable to issue passes to visitors in order to maintain control throughout the facility. There is no general agreement on the need for such passes during regular visiting hours, but the hospitals which have adopted the policy have generally been pleased with the increased efficiency in traffic control they achieve without materially interfering with those who are visiting patients.

It is important to create an effective balance between security needs and a concern for the patient, whose recovery can be speeded, or at least whose stay can be made more pleasant, by visits from friends and family.

Since there is the likelihood of a brief jam-up in the lobby at the beginning of visiting hours in any hospital where the use of visitor's passes is a policy, it is desirable to work out a pass design and issuance procedure that can be handled with the greatest speed and efficiency and that will cause the least inconvenience to the visitor.

A number of different visitor pass systems are in use in hospitals around the country, but a general profile would include many or all of the following features:

1) Color coding by floor or department to be visited.

2) Two cards made up for each bed.

3) Both cards clearly marked with name and address of hospital, room number and bed assignment, and a recap of visiting hours. Special color-coded cards may be issued for other than regular visiting hours authorization.

4) Patient's name can be written in pencil on ½-inch masking tape attached to the two passes assigned to that bed. When that patient leaves, his name can be erased or the masking tape replaced with another name.

5) The reverse side of the card pre-addressed for mailing with return postage guaranteed.

6) The pass of sufficient size to make it too bulky to put easily in handbags or pockets, thus discouraging visitors from taking it home with them after the visit.

Such a system has certain disadvantages, in that some visitors and patients may feel that the procedure is too restrictive, since only two visitors per bed at any one time are permitted. There is also the problem already mentioned of lines forming, with some inevitable delays at the start of visiting hours, particularly in the early evening.

The positive aspect of these controls will probably outweigh the disadvantages, however. Floor traffic in patient areas is diminished. Doctors and nurses alike will find conditions for treatment of patients are considerably improved with the reduction of milling visitors. These benefits accrue from a traffic control system that has as its overall effect the reduction of thefts from patients and from the hospital by otherwise unidentified opportunists.

INTERNAL CONTROLS

Employee Screening

As in every business, it is important for a hospital to take pains to check into the background of every prospective employee. It goes without saying that applicants with a background of offenses involving narcotics, sex or crimes of violence would be poor prospects for employment in a patient care facility. This would probably also be true of those having a long and continuing history of petty offenses in any area. Such records can only be uncovered by fingerprinting all applicants and checking them through police files.

Many hospitals additionally require all employees to submit to package inspections, and to consent to undergo a polygraph examination (in those states where such examination may legally be required) as a condition of employment. Since polygraph examinations are expensive, they are rarely undertaken as a part of hospital hiring practices, but the very fact that the applicant must consent to such a procedure will discourage a significant number of undesirables from following up on their pursuit of employment at the hospital.

Services, Cash and Supplies

Control of services, cash and supplies is as important in a hospital as in any business. This is harder to manage because of the frequent emergency nature of the services and supplies provided under circumstances that are often totally alien to any consideration of remuneration or cash recovery. Certainly a dedicated medical staff is concerned with the immediate welfare of the patients and not with the administrative routine of debts and credits.

Unfortunately, every hospital must deal with the cold realities of the economies governing its continued operation—its very existence. To do this it must establish controls over all its assets, including its accounts receivable. These safeguards must begin with a well-planned system in which efficient procedures will include painstaking record keeping, properly audited by precise accounting practices.

Medication and Services

Since the principal commodities supplied by a hospital are services and medications, controls must be established over all charges for them. Every time a charge is overlooked or is lost in the urgency of the moment, income is reduced and expenses are increased. Multiplied many times, these omissions can cost the hospital enormous amounts of money in a very short time.

Many different systems have been adopted to ensure the recording of all medication or service. Most of them essentially involve a system of dual entries, which are then cross-checked upon the patient's discharge. In its simplest form this involves assigning a slip with adequate carbons to each patient when he checks into the facility. This slip accompanies the patient throughout his stay in the hospital. Any charges incurred in his treatment are entered on his slip, and a duplicate charge is entered into the appropriate record of the doctor, nurse, technician, or other responsible personnel providing the service. These records are then checked against the patient slip upon the patient's discharge.

However such a system is implemented, it can never be totally effective without the full cooperation of all personnel involved. Medical personnel, especially in an emergency situation, frequently find that financial matters are the last thing to consider when they are confronted with circumstances involving great pain or even the survival of the patient. However commendable this humane attitude may be, there must be an orderly system of financial accountability or the hospital would soon be unable to provide those services so necessary to the welfare of the community.

Periodic checking of departmental records against individual patient slips will go a long way toward eliminating loss by omission and will identify those of the staff who are most inclined to this oversight. With the high cost of medical care today, it is the obligation of every staff member to keep general expenses at their lowest possible level. An essential step in this direction is the assignment of fees and charges to those individuals requiring specific services, and preventing such expenses from becoming general, unassigned overhead which every patient, without discrimination, must help to defray.

Cash Control

Needless to say, every hospital must maintain the strictest security in situations where cash is handled. Safeguards in those areas must be as exacting in a hospital as they would be in any other business.

This is equally true in the receiving and inventory of goods, the disbursements of accounts payable, and the handling of all accounts receivable. Without an adequate system that separates these functions and pinpoints accountability, a hospital is precisely as vulnerable to internal theft, embezzlement or collusion as any business of any kind.

In the same sense, a hospital should follow sound business practice in having regular audits of its books by an outside accounting firm. In the long run this is money wisely spent and could, in fact, result in savings many times the cost of such audits.

Insurance

Insurance is another factor which is as important in the asset protection program of a hospital as elsewhere. Property damage resulting from fire, flood, wind, hail, etc., can destroy a hospital's assets more effectively than the most accomplished thief. When the internal and external assaults upon hospital property are considered along with the potential for natural disaster, we can see an imposing array of dangers that must be dealt with in protecting the facility. Adequate insurance, based upon replacement value and attendant costs of temporary relocation, is essential to an overall protection plan.

FIRE

No occupancy is entirely free from the threat of fire. Hospitals are by no means an exception. In fact, Stephen Barlay, in his disturbing book *Fire,* points out that, statistically, the chances of fire occurring in a hospital are one hundred times greater than in any other building in New York.[19]

This threat to a hospital is particularly severe because many of its occupants are in such a condition that they are unable to fend for themselves in an emergency. In addition, few of the patients, even those who are well on the way to convalescence, are familiar with the hospital beyond their own limited area. If they were forced to make a rapid evacuation, they would have a difficult and dangerous trip through totally unfamiliar territory. The potential for disaster is obvious.

Prevention of Fire

Although the major ignition source of hospital fires (as well as those in most other buildings) is electrical equipment, with cigarettes and matches close behind, it is in the various fuel sources that hospitals are unique. Oxygen, anesthetics, volatile liquids of various kinds—all can be dangerous under the proper circumstances.

Flammable Waste

In addition, Mr. Barlay points out that, whereas the average American office building generates about two pounds of waste paper per hundred square feet of floor space a day, or several tons a week for the entire building, waste material in a hospital averages about eight pounds per bed each day.[20] With offices and laboratories included, the total waste generated by a thousand-bed hospital would amount to about four tons each day.

This waste is highly flammable. With this amount of fuel generated daily, there is a very real hazard unless the tightest controls are maintained over the movement, holding and collection of all rubbish. Maintenance personnel and other staff members must be taught to deposit all waste material in metal containers, which must be emptied at least once daily. Trash outside the hospital awaiting pick-up must be held in metal containers sufficiently far from buildings so that they will not represent a hazard.

Flammable Liquids

Flammable liquids, which pose a special problem to hospital safety, should be stored, dispensed and disposed of in appropriate safety containers and storage cabinets. Flammable liquids with a flash point above 20°F. must be handled under the strictest supervision and should be limited in the quantity permitted on the premises at any one time.

Electrical Hazards

Electrical wiring must be inspected regularly for any signs of wear or malfunction. Normally extension cords should be run overhead—avoiding any contact with nails, pipes, hooks, grilles or other conductors. In cases where extensions must run along the floor, they must be heavily insulated and protected against the water with which they would come in contact when the floors are scrubbed.

Vulnerable Areas

Elevator and dumbwaiter shafts must be cleaned periodically to prevent the inevitable build-up of oil, grease and rubbish, which can create stubborn and dangerous fires that are difficult to combat. Linen and rubbish chutes should be washed weekly or more to prevent dangerous dust hazards.

Operating and delivery-room floors should be tested regularly for static electricity. The importance of such testing cannot be overstated. Even though today less than twenty per cent of the anesthetics in use are flammable, accidents can occur. A spark from static electricity in an atmosphere containing cyclopropane or ether—both of which are still in limited use as anesthetics—can be deadly.

In the November, 1969, issue of the *Fire Journal* of the National Fire Protection Association, Dr. James M. McCormick points out that, while pre-1941 reports suggest that the incidence of fire and explosion was in the area of 1 in 300,000 anesthetics, it would now appear that such incidence is in the range of 1 in 600,000 anesthetics. He goes on to indicate that it is impossible to determine the actual frequency because medical reporting of such cases is inhibited by adverse publicity and the fear of malpractice liability. In any event, it is a very real danger and one that must be dealt with by the hospital.

Lavatories must be cleaned regularly—not only for purposes of sanitation but to avoid the accumulation of refuse in those areas. Overstuffed waste baskets are always a hazard and must not be permitted. Since approximately twenty per cent of all hospital fires are caused by smoking and careless use of matches, any area where paper might accumulate is a potential danger. Fireproof receptacles for paper can provide an extra margin of safety in lavatories or other areas where combustible refuse might accumulate.

Public areas such as lounges, cafeterias, waiting rooms or lobbies must be amply supplied with appropriate receptacles for use of smokers. As long as smoking is permitted in areas where people congregate, there is the danger of fire. Fires from a cigarette or cigar butt are particularly treacherous, since it may smoulder in a cushion or on a rug under a couch and not become evident until long after the area is deserted. It is advisable to supply containers of sand (which are regularly cleaned of gum wrappers and other scraps of paper that find their way into them) and ashtrays of adequate size which are equipped with grooves or snuffers long enough to hold the entire cigarette safely.

Inspection

It should be the responsibility of the Security Director to make a daily inspection of the facility to determine the effectiveness of fire prevention procedures. He should assure himself that all specified inspections have been made and that all directives relative to fire safety have been followed out.

- Extinguishers, fire hose, and the water supply for fire lines must be checked regularly.
- All oxygen shutoff valves must be checked and tagged.
- Exit signs must be checked regularly for visibility and for burnt-out bulbs.
- Alarm systems must be checked regularly.
- Signs designating no smoking or special hazard areas must be checked for accuracy and applicability as the needs for them arise or dissipate.
- Fire stairs or smoke towers must be checked for any obstructions or any other factors which could possibly impede a smooth and orderly evacuation if the need should arise.

All of the above procedures must be observed in order to establish the beginnings of a fire prevention program. Obviously, each hospital has its own problems. They must be dealt with after a careful inspection and evaluation of its own circumstances.

Unfortunately, many hospitals are still being built without adequate thought being given to lessons that might have been learned from tragic fires of recent history. Too many hospitals exist with inadequate or badly planned exit arrangements, or with highly combustible interiors and inadequate fire-fighting equipment. In such a facility every effort must be made to correct or overcome these inadequacies before a tragedy occurs. Every administrator must be made aware of these dangers, and a plan must be devised to overcome them. In most cases, it will be the responsibility of the Security Director to so advise the hospital administrator.

Emergency Exits

As a measure of the adequacy of the exits in a hospital, we might refer to the National Fire Protection Association's *Life Safety Code,* which indicates the types of permissible exits and their capacity:

Types of Exits	Maximum Capacity Per Unit
Doors directly outside	30
Stairs and smokeproof towers	22
Ramps	30
Horizontal exits	30
Outside stairs	22

The *Code* further specifies that each floor or fire section of a hospital have a minimum of two exits, remote from each other, one of which must be a door directly to the outside or a stair or a smokeproof tower.

Access to these exits is summarized in the following table:

Travel distance to exits	100 feet from the door of any room or 150 feet from the farthest point in any room
Minimum width of access to exits	8 feet
Maximum dead end in corridor	30 feet
Minimum width of doorways from sleeping rooms and diagnostic and treatment areas to exits	44 inches

Minimum width of
  doorways to nurseries   36 inches

The capacity of exits in hospitals is considerably lower than in other types of occupancy. Not only can hospital patients be expected to move more slowly than others, but it can also be expected that some evacuees will be in beds or on stretchers, which will considerably reduce the movement possible in a given period of time.

A careful study of the NFPA's *Life Safety Code* with regard to institutional occupancies is an essential part of the providing of adequate fire protection to a hospital. Since many hospitals fail to conform to the *Code* in some respects, it is particularly important to be aware of those shortcomings and to make provisions to accommodate those needs.

Supervision of the Fire Program

In this section on fire prevention and protection in a hospital, we have assumed that the supervisor of this function was the Security Director. This is generally the case. Many hospitals, however, have employed full-time fire marshals who report to the administrator and are not involved in or answerable to Security in other respects. In other situations the responsibility for fire safety is assigned to the engineer. While many engineers are or become highly proficient in fire safety, just as many are not; and they are often too busy with their other duties to become even routinely knowledgeable.

It would appear that the most satisfactory arrangement is to assign the total job of protection to the Security Department. In this way the problems of fire, violence, theft, safety, disaster control and security services are integrated into a broad program, each element of which compliments and supports the other. Obviously experienced personnel from city fire departments can be of inestimable value as members of such a total protection program, but it could be argued that their value would be diminished if their efforts were isolated from the total security program.

Fire Marshal

No matter where he fits into the organizational chart or whatever his background, however, every hospital must have someone who acts as Fire Marshal. He will be directly responsible for developing training programs, fire reporting systems, and evacuation procedures; and establishing and administering the vital area of fire prevention. He will establish and maintain liaison with the local fire department, and he will be responsible for keeping abreast of the latest developments in every aspect of fire safety. He will assume complete charge during a fire emergency until the arrival of the fire department. In a large hospital complex he will undoubtedly be a very busy man.

Fire Alarm Systems

The first step in the marshal's plan is to survey the area minutely and to pinpoint problems peculiar to specific areas. He must next set up a fire reporting system by alarm or phone or both. This system must report the fire and its location to the Fire Marshal and to the city fire department; and it must be announced in some way over the hospital paging system so that all personnel can respond appropriately.

The usual alarm system in larger hospital complexes is set up to sound various combinations of bells, each combination coded to indicate a different location in the facility. Phones throughout the hospital can also be used to call in the location of a fire. The paging system should be used to announce the fire and its location. Most hospitals use a code name for a fire condition to avoid alarming visitors and patients. In the event the fire becomes more serious and requires some evacuation, the condition might then be announced in plain talk over the system.

Fire Planning and Training

The next step in the fire safety program is the establishment of a plan for various fire situations. Everyone in the fire area must have specific assigned duties which are to be carried out under the direction of floor captains.

In order for such a plan to work at all effectively, all personnel must receive fire training as part of their orientation. This includes chief residents, interns and residents. Each employee must know how and where to turn in an alarm; he must know the location and the proper use of fire extinguishers and fire hoses; he must know the location of *all* exits and be familiar with the alternative actions necessary if any of them are blocked. He must be familiar with the potential fire hazards, his specific duties in given situations, and evacuation procedures. Above all, he must be thoroughly indoctrinated in the need for immediate action. He must be persuaded that prompt and proper fire fighting immediately after reporting the fire is essential. He must be taught that the first two or three minutes of a fire are the most important—that it is in this brief period that training pays off or tragedy can occur.

Only frequent fire drills can drive these lessons home. For these drills to be fully effective they must be observed, and the performance of each participant must be evaluated. Through the repetition of drills and the follow-up critique, nurses, interns and hospital employees can become an effective emergency team.

It is important to remember, however, that this training program cannot be a routine, *pro forma* performance of some sort of a charade. A hospital staff consists of many dedicated, hardworking professionals whose focus is on service to the patients. Unless their training in fire safety is presented in an engaging, perhaps dramatic way, they may well overlook its significance and its immediacy, and treat it simply as a necessary but unpleasant chore. If that is their reaction, the training program must be re-thought and re-designed at once for the safety of everyone in the facility.

Training in fire prevention never ends. New employees need full indoctrination and old ones need constant review of fire emergency procedure. Only by effective training can a hospital mobilize its staff into an efficient force for safety.

Chapter 10

CARGO
SECURITY

DIMENSIONS OF THE PROBLEM

During the three years of hearings by the Senate Select Committee on Small Business (1970-1972), testimony of witnesses and staff investigation led to the conclusion that the direct losses from theft in cargo carried by the air, truck, rail and maritime carriers amounted to almost $1-1/2 billion in 1970. And that figure seems to be increasing rapidly.

The indirect costs resulting from claims processing, capital tied up in claims and litigation, and market losses due both to non-delivery and underground competition from stolen goods, are estimated at between two to seven dollars for every dollar of direct loss—an eight to ten billion dollar annual loss in the national economy.

Inadequacy of Security

The shipment of goods is vital to the economy and ultimately to the survival of the country. Yet the Committee found that law enforcement efforts, conducted by local and state police and the FBI, were totally inadequate to prevent this enormous theft of cargo and to apprehend the criminals involved.

Testimony additionally indicated that, although some trucking concerns were making an effort to establish effective security programs, a considerably greater part of the industry, from shippers to receivers, from warehouse managers to drivers, were bogged down in confusion and ineffective half-measures that served more to nurture and encourage crime than to prevent it. Control systems were frequently misunderstood, paperwork was complicated and often unchecked, physical security was inadequate, supervision was inadequate, traffic control was poor, and security policies were often misdirected or unenforced.

The Committee called for a program directed at crime *prevention* and law enforcement *coordination* rather than an expansion of police or reactive forces. The Committee concluded that security measures do exist which can halt the increase in theft in the shipping industry, and it encouraged the industry, along with law enforcement officials, to set up coordinated and cooperative programs that will apply the most advanced technology to the solution of these problems.

The Role of Private Security

Since, according to an analysis made by the Department of Transportation, 85 percent of goods and materials stolen go out the front gates on persons and vehicles which have been authorized to be in cargo handling areas of transportation facilities, it would appear that by far the greater burden falls upon the security apparatus of the various private concerns involved. It is true that public law enforcement agencies must make a greater effort to break up organized fencing and hijacking operations, and they must find a way to cut through their jurisdictional confusions and establish more effective means of exchanging information; but the bulk of the problem lies in the systems now employed to secure goods in transit.

There is no universally applicable solution to this problem. Every warehouse, terminal and means of shipment has its own particular peculiarity. Each one has weaknesses somewhere, but certain principles of cargo security, when thoughtfully applied and vigorously administered, can substantially reduce the enormous losses which are so prevalent in today's beleaguered transport industry.

Importance of Accountability

The paramount principle is accountability. Every shipment, whatever its nature, must be identified, accounted for, and accounted to some responsible person at every step in its movement. This is difficult in that the goods are in motion and there are frequent changes in accountability, but this is the essence of the problem. Techniques must be developed to refine the process of accountability of all merchandise in shipment.

Control Procedures

This accountability must start from the moment an order is received by the shipper. As an example of a typical controlled situation, we might refer to a firm that supplies its salesmen with sales slips or invoices that are numbered in order. This is very important if any control is to be maintained over these important forms. Without numbering, sales slips could be destroyed and the cash, if any, could be pocketed; or they could be lost so that the customer might never be billed. When these invoice forms are numbered, they can be charged out to the salesman, in which case every invoice can and should be accounted for. Even those forms which have been spoiled by erasures or physical damage should be voided and returned to billing.

Merchandise should only be authorized for shipment to a customer on the basis of the regular invoice form. This form is filled out by the salesman receiving the order and sent to the warehouse or shipping department. The shipping clerk signs one copy, signifying that the order has been complied with, and sends it to accounts receivable for billing purposes. The customer signs a copy of the invoice indicating receipt of the merchandise, and this copy is returned to accounts receivable. It is further advisable to have the driver sign the shipping clerk's copy as a receipt for his load. In some systems a copy of the invoice is also sent directly to an inventory file for purposes of on-going inventory and audit. Returned merchandise is handled in the same way, but in reverse.

In this simplified system there is a continuing accountability for the merchandise. If anything is missing or unaccounted for at any point, the means exist whereby the responsibility can be located. Such a system can only be effective if all numbered invoices are strictly accounted for, and where merchandise is assembled and shipped only on the basis of such an invoice. The temptation to circumvent the paper work in the name of "rush order" or "emergency" frequently arises, but if the company succumbs, losses in embezzlement, theft or lost billing can be substantial.

Similarly, transport companies and freight terminal operators must insist on full and uninterrupted accounting for the goods in their care at every phase of the operation from shipper to customer.

Separation of Functions

Such a system can be compromised by collusion between the various people who constitute the links in the chain leading from order to delivery, unless efforts are made to establish a routine of regular, unscheduled inspections of the operations down the line; and, depending on the nature of the operation, regular inventories and audits.

It is advisable, for example, for the shipper to separate the functions of selecting the merchandise from stock from the packing and loading function. This will not in itself eliminate the possibility of collusion, but it will provide an extra check on the accuracy of the shipment. And as a general rule, the more people (up to a point) charged with the responsibility and held accountable for merchandise, the more difficult and complex a collusive effort becomes.

In order to clearly fix accountability, it is advisable to require each person who is at any time responsible for the selecting, handling, loading or checking of goods to sign or initial the shipping ticket which is passed along with the consignment. In this way errors, which are unfortunately inevitable, can be assigned to the man responsible. Obviously any disproportionate number of errors traced to any one person or to any one aspect of the shipping operation should be investigated and dealt with promptly.

Similarly, at the receiving end the ticket should be delivered to a receiving clerk. The truck will then be unloaded by appropriate personnel who will verify the count of merchandise received without having seen the ticket in advance. In this way each shipment can be verified without the carelessness that so frequently accompanies a perfunctory count that comes with the expectation of receiving a certain amount as specified by the ticket. it will also tend to eliminate theft in the case of an accidental or even an intentional overage. Since the checker doesn't know what the shipment is supposed to contain, he cannot readily rig the count.

Driver Loading

Many companies, with otherwise adequate accountability procedures, permit drivers to load their own trucks when taking a shipment. This practice can defeat any system of theft prevention, since the driver is accountable only to himself when such a method of loading is in effect. The practice should never be condoned. In the long run it would not be an economy in extra time spent in checking the load or extra personnel needed to do it. The potential losses in theft of merchandise by overloading or future claims of short deliveries would almost certainly exceed the costs involved in the time or personnel involved in instituting sensible supervisory procedures.

KINDS OF THEFT

Pilferage

According to the Department of Transportation analysis referred to above, 25 percent of the cargo theft in this country is the result of pilferage or thefts of less than a case. Thefts of this nature are generally held to be impulsive acts, committed by persons operating alone and who pick up an item or two of merchandise which is readily available when there is small risk of detection. Normally the pilferer takes such items for his personal use rather than for resale, since most of those who are termed pilferers are unsystematic, not of a committed criminal nature, and are unfamiliar with the highly organized fencing operations that could readily dispose of such loot.

Forms of Pilferage

Such pilferage is always difficult to detect. Because it is a crime of opportunity, it is rarely committed under controlled circumstances which can either pinpoint the culprit or gather evidence that would later lead to his discovery. Normally the items taken are small and readily concealed on the person or easy to transport and conceal in a car.

Pilferage is usually aimed at items in a freight terminal which are awaiting transshipment. In such instances merchandise may be left unprotected on pallets, hand carts or dollies awaiting the arrival of the next transport. In this mode it is highly susceptible to pilferage as well as to a more organized plan of theft.

Broken or damaged cases offer an open invitation to pilferage if supervisory or security personnel fail to take immediate action. Accidental dropping of cases to break them open is a common device used to get at the merchandise they contain. If each such case is carefully logged, listing the name of the man responsible for the damage as well as the names of those who instantly gather at the scene, a pattern may emerge that will enable management to take appropriate action.

Whatever form pilferage takes, it can be extremely costly. Although each individual instance of such theft may be relatively unimportant, the cumulative effect can be enormous. 25 percent of an estimated total direct loss of $1½ billion adds up to a staggering bill for petty theft.

Accountability

Here again, accountability controls can provide an important deterrent to this kind of loss. If some one person is responsible for merchandise at every stage of movement or storage, the feasibility of this kind of theft can be substantially reduced.

In those cases where they do occur, a rapid and accurate account of the nature and extent of such losses can be an invaluable tool in indicating the corrective action to be taken. Properly supervised accountability controls can locate the point along the handling process where losses occurred and, even in those cases where they will not identify the culprit, they will underscore any weaknesses in the system and indicate trouble areas which may need more or different security application.

Movement of Personnel

Movement of personnel in cargo areas must be strictly controlled. All parcels must be subject to inspection at a gate or control point at the entrance to the facility. Private automobiles must be parked outside the area immediately encompassing the facility and beyond the check point. All automobiles should be subject to inspection upon departure if a parking area is provided.

Every effort should be made to keep employee morale high in the face of such security efforts. Though some managements have expressed an uneasiness about inspections and strict accountability procedures, fearing that they might damage company morale, it should be pointed out that educational programs aimed at acquainting employees with the problems of theft and stressing everyone's role in successful security have resulted in boosting morale and in enlisting the aid of all employees in the effort.

Here, as in other areas of security, employees should be encouraged to report losses immediately. They must never be encouraged to act as informers or asked to report on their co-workers. If they simply report the circumstances of loss, it is the job of security to carry forward such investigation or to take such action as may seem indicated.

Theft

Referring again to the DOT report, we find that 60 percent of the theft of cargo consists of merchandise in quantities of one or more cases but less than a full load. Thefts in such amounts are no longer in the category of pilferage. This becomes thievery, usually engaged in by one or more persons who are in it for profit—for resale through traffickers in stolen goods.

The thief may or may not be an employee, but since in either case he needs information about the nature of the merchandise on hand or expected, he will usually find an accomplice inside who is in a position to have the information. He is interested in knowing what kinds of cargo are available in order to make a decision what merchandise to hit, depending on its value and on the demands of the fencing organization with which he deals.

Dealers in stolen goods are subject to the vagaries of the marketplace in the same way legitimate businessmen are; whereas a certain kind of merchandise may find a ready market today, it may move slowly tomorrow. Such dealers are anxious to move their goods rapidly, not so much because of fear of detection (since even if they are found, mass produced goods of any nature are difficult if not impossible to identify as stolen once they find their way into other hands), but because they generally want to avoid the overhead and the attention created by a large warehousing operation.

Removal of Goods

Once the thief has the information, he needs to arrange to take over the merchandise and remove it from the premises. To do this he will usually try to work with some employee of the warehouse or freight terminal he has singled out.

In cases where accountability procedures are weak or inadequately supervised, he has few problems. In tighter operations he may try to bribe a guard or he might forge papers which represent him as an employee of a customer firm. He might even create confusion, such as a fire in a waste bin or a broken water pipe, in order to divert attention from his actions in those few moments he needs to accomplish the actual theft. Generally the stolen goods are taken from the facility in an authorized vehicle driven either by the thief, who has supplied himself with false identification papers and forged shipping documents, or by an authorized driver who is, more often than not, working with him.

Disposal of Stolen Goods

Disposal of the goods usually presents no problem, since it is customary for him to steal on an order placed for any one of certain kinds of merchandise. His loot is normally pre-sold.

It might be noted at this point that it is for this reason principally that cooperation between private security and public law enforcement is so vital in the war against cargo thefts. No thief will continue in his efforts unless he has a ready market for the goods he steals. In the hearings of the Select Committee on Small Business, one of the most inescapable conclusions to be drawn was that the fences were the kingpins behind the majority of the thefts taking place. It is they who dictate the nature of the merchandise to be taken, the price to be paid, and the amount wanted. They direct the thefts without ever becoming involved—in many cases never even seeing the merchandise they have bought, warehoused and sold. Without the fences—many of whom are otherwise legitimate businessmen—the markets would shrink, the distribution networks would disappear, and losses would be dramatically reduced.

Unfortunately, only sporadic and ineffective efforts have been made to break up the big fencing operations, and the business of thievery thrives. Any assistance that private security can give to public law enforcement by way of instant and full reports on thefts could help to combat these shady operations, and all private industry will benefit immeasurably in the long run.

Terminal Operations

Terminal operations are probably more vulnerable to theft than other elements of the shipping system. Truck drivers mingle freely with personnel of the facility, and associations can readily develop that often lead to collusion. Receiving clerks can receipt goods that never arrive; shipping clerks can falsify invoices; checkers can overload trucks, leaving a substantial percentage of the load unaccounted for and therefore disposable at the driver's discretion.

Here again, these thefts can be controlled with a tight accountability system, but too often such facilities fail to install such a procedure or to follow up on it after it is in effect. This is poor economy, even under the most difficult situations when seasonal pressures are at their highest.

Railway employees on switching duty at a freight terminal can also divert huge amounts of goods. They can easily divert a car to a siding accessible to thieves who will unload it at an opportune time. This same device can be employed to loot trucks which have been loaded for departure the next morning. Unless these trucks are securely locked and parked where they can be under surveillance by security personnel, they can be looted with ease. Drivers can park their trucks unlocked near a perimeter fence for later unloading, unless the positioning and securing of the vehicle is properly supervised.

In all cases a professional thief is a man with a mission. He cannot be deterred by the threat of possible detection. He recognizes the possibility, accepts the risk, and makes it his business to circumvent detection. Only alert and active countermeasures will serve to reduce losses from his efforts.

Surveillance

There must be a strict guard surveillance at entrances and exits, and there must be patrol activity at perimeters, and through yards, docks and buildings. Key control must be tight and painstakingly supervised. Cargo should be stored in controlled security areas which are enclosed, alarmed, and burglar-resistant. High value cargo should be stored in high security areas within the cargo area. Special locks, alarms and procedures governing access should be employed to provide the highest possible security for these sensitive goods.

Employee Screening and Controls

Employees must be carefully screened. Screening is a prudent practice in any operation, but particularly so in a terminal where the thief is in such constant, intimate contact with such a tempt-

272 INTRODUCTION TO SECURITY

ing variety of merchandise. Employment in such a facility is the goal of many thieves, but a reasonably thorough screening procedure will weed out many of them.

Every effort must be made to frustrate the thief who may have gotten through the employee screening procedures in order to work out a schedule of theft from the inside. His efforts can be counteracted to some degree by denying him the information he needs to plan for major attacks. Shipments of unusual value should be confidential, and only those employees who are directly concerned with loading or unloading or transporting such shipments should be aware of their schedules. Teletype information about such movements should be restricted, and trailer numbers should be covered while the vehicle is in the terminal. Employees involved in any way with such shipments should be specially selected and further indoctrinated in the need for discretion and confidentiality.

A Total Program

Adequate physical security installations supported by guard and alarm surveillance will go a long way toward protecting the facility from the thief, but these measures must be backed up by proper personnel and cargo movement systems, strict accountability procedures, and a continuing management supervision and presence to insure that all systems are carried out to the letter. Regular inspections of all facets of the operation, followed by prompt remedial action if necessary, are essential to the success of the security effort.

PLANNING FOR SECURITY

It is important to the security of any transportation company, shipper or freight terminal operation to draw up an effective plan of action to provide for overall protection of assets. The plan must be an integrated whole wherein all the various aspects are mutually supportive.

In the same sense, in large terminal facilities occupied by a number of different companies, all individual security plans must be integrated to provide for overall security as well as for protection of individual enterprises. Without full cooperation and coordination among participating companies, much effort will be expended uselessly and the security of the entire operation could be threatened.

Such a plan should establish area security classifications. Designated parts of the building or the total yard area of the facility should be broken down into controlled areas, limited areas and exclusion areas. These designations are useful in defining the use of specific areas and the mounting security classifications of each of them.

Controlled Areas

Controlled areas are those areas whose access is restricted as to entrance or movement by all but authorized personnel and vehicles. Only part of a facility will be designated a controlled area, since general offices, freight receiving, personnel, rest rooms, cafeterias, locker rooms, etc., may be used by all personnel, some of whom would be excluded if these facilities were located within an area where traffic was limited. Within this area itself, all movement should be controlled and placed under surveillance at all times. It should additionally be marked by a fence or other barrier, and access to it should be limited to as few gates as possible.

Limited areas are those within the controlled area where a greater degree of security is required. Sorting, handling of broken lots, storage, and recoopering of cases might be vulnerable functions handled in these areas.

Exclusion areas are used only for the handling and storage of high value cargo. They normally consist of a crib, vault, cage or room within the limited area. The number of people authorized to enter this area should be strictly limited, and the area should be under surveillance at all times.

Since such areas should be locked whenever they are not actually in use, careful key control is of extreme importance.

Pass Systems

All employees entering or leaving should be identified and checked for their authorization to be there. Each employee should be identified by a badge or pass, using one of several systems:

- The single pass system, wherein the badge or pass coded for authorization to enter specific areas is issued to an employee who keeps it in his possession until his authorization is changed or until he terminates.

- The pass exchange system, in which he exchanges one color-coded pass at the entrance to the controlled area for another which carries a different color code specifying the limitations of his authorization. Upon leaving he surrenders his controlled area badge in exchange for his basic authorization identification. (In this system the second badge never leaves the controlled area, thus reducing the possibility of switching, forging or altering.)

- The multiple pass system, which provides an extra measure of security by requiring that an exchange take place at the entrance to each restricted area within the controlled area.

Vehicle Control

The control of the movement and the contents of all vehicles entering or leaving a controlled area is essential to the security plan. All facility vehicles should be logged in and out on those relatively rare occasions when it is necessary for them to leave the controlled area. They should be inspected for load and authorization.

All vehicles entering the controlled area should be logged and checked for proper documents. The fastest and most efficient means of recording necessary data is by a camera such as a Regiscope, which will record all of the required information on a single photograph. This should include the driver's license, truck registration, trailer or container number, company name, way bill number, delivery notice, a document used to authorize pickup or delivery, time of check, and the driver's picture if it is not on his license.

The seal on inbound loaded trailers should be checked and the driver issued a pass which will be time-stamped on entering and leaving the area and which will designate the place for pickup or delivery.

All vehicles leaving the area should surrender their pass at the gate where their seals will be checked against shipping documents. Unsealed vehicles will be inspected, as will the cabs of all carriers leaving the facility. Partial load vehicles should be returned to the dock from time to time on a random basis for unloading and checking cargo under security supervision.

Loading and unloading must be carefully and constantly supervised, since it is generally agreed that the greater part of cargo loss occurs during this operation and during the daylight hours.

Other Security Planning

The security plan must also specify those persons having access to security areas.

It must specify the various components necessary for physical security such as barriers, lighting, alarm systems, fire protection systems, locks and communications.

The plan must detail full instructions for the guard force. These must contain general orders applicable to all guards and special orders pertaining to specific posts, patrols and areas.

There must be provision for emergency situations. Specific plans for fire, flood, storm or power failure should be a part of the overall plan of action.

After the security plan has been formulated and implemented, it must be re-examined periodically for flaws, for ways to improve it and to keep it current with existing needs. Circulation of the plan should be limited and controlled. And as a final note of emphasis, it must be remembered that such a plan, however well conceived, is doomed at the outset unless it is constantly and carefully supervised.

CARGO IN TRANSIT

The Threat of Hijacking

Although the crime itself is dramatic and receives much publicity, armed hijacking of an entire tractor-trailer with a full load of merchandise represents only 10 percent of the losses suffered by the shipping industry as a whole. This is clearly not to say that it is a minor matter. On the contrary, such a crime is of extreme importance to the carrier taking such a loss because the enormity of the theft represents a huge financial blow in one stroke, whatever its significance in the overall percentages.

There is, however, little that can be done by a driver who is forced over by a car carrying armed and threatening hijackers.

When it comes to that point, he has little choice but to comply. The load is lost but hopefully the driver is unhurt. There is little that private security can do in such cases, and it is indeed fortunate that the incidence of such crimes is as low as it is. The matter is in the hands of public law enforcement agencies and must be handled by them.

If there is cause to believe that hijacking of a load is an imminent possibility, trucks should be scheduled for non-stop hauls and re-routed around high risk areas. Schedules should be adjusted so that carriers do not pass through high risk areas at night. In extreme cases, trucks might be assigned to travel in pairs or in larger convoys. Very high value loads that are deemed especially vulnerable can be followed by company cars. Such procedures constitute selective protection at best, since they can be used only infrequently and are impractical for general application.

Other aspects of theft on a line haul can be dealt with, however, and it is important that procedures be established that will serve to protect the cargo.

Personnel Qualifications

The cardinal rule in the management of a trucking concern is that those assigned to line haul duties must be of the highest integrity. Drivers and helpers must be carefully screened before they are hired and they must be carefully evaluated by personnel and security managers before they are given this critical responsibility. An irresponsible or criminally inclined driver can cost the company all or part of his load, and whether the loss is unintentional or the result of the driver's carelessness, the cost to the company will be the same.

Many cases have been reported where the driver has set himself up as a "victim" of a hijack. This is difficult if not impossible to prevent unless the honesty of the driver is unquestionable—an attitude which can only be determined by a careful screening process and regular analysis of the man's behavior and performance. Any changes in demeanor or in his life style should be noted as a possible deterioration of morale or a basic change in attitude toward his job that might lead to future problems.

Procedures on the Road

All employees should receive specific instructions about procedures to be followed in every predictable situation on the road. The vehicle should be parked in well-lighted areas where it can be observed. It should be locked at all times, even if the driver sleeps in the cab. Trailers should be padlocked as well as sealed.

Drivers must be instructed never to discuss the nature of their cargo with anyone. Thieves frequently hang out at truck stops hoping to pick up information about the nature of loads passing through. All too often the most innocent conversations among truckers can lead to the identification of a trailer containing high value merchandise which, when spotted, becomes a target.

The driver should never deviate from the preplanned route. In case he is forced to take a detour or his rig breaks down, or if in any way his schedule cannot be met, he must notify the nearest terminal immediately.

Trucks should be painted on the top and sides to facilitate identification by helicopter in the event of theft.

Seals

Among the many seals available today, the one in most common use is the metal railroad type. This is a thin band of metal which is placed and secured on the trailer in such a manner that the door or doors cannot be opened without breaking it, thus revealing that the doors have been opened and a theft has or may have taken place. They are easy to break and must be broken when the destination has been reached and the merchandise is unloaded. They in no way secure the doors, but are placed there simply as a device to indicate whether the doors have been opened at any point between terminal stops. Each seal is numbered and should identify the organization which placed the seal.

All doors on a trailer must be sealed. Those trucks or trailers with multiple doors may habitually load by only one, in which case those doors not in regular use may carry the same seal for months at a time with only the rear being regularly sealed and unsealed.

Seal numbers must be recorded in a permanent log as well as in the shipping papers. The dock superintendent or security personnel should be responsible for recording seal numbers and affixing seals on all trucks. The seals should be positioned in such a way that locking handles securing the door cannot be operated without actually breaking the seal. Some truck or railway car locking devices are so large that several seals may have to be used in a chain to properly seal the carrier. In this case all seal numbers must be recorded.

Re-Sealing

Trailers loaded to make several deliveries along the route must be resealed after each stop. To accomplish this, enough seals must be issued to be placed on the truck after each delivery is made. In this case the truck is sealed at the point of origination and the seal number logged and entered in shipping documents. The additional, as yet unused, seals are also logged, and the numbers to be used at designated points of delivery are entered in the shipping documents.

When the first point is reached, the receiving clerk verifies the seal number of the truck on arrival. After his merchandise is unloaded, he affixes the next seal as directed by the shipping documents. This procedure is followed at all stops, including the last one. There a seal is affixed to the now empty vehicle, which may return to the point of origin, where the seal will again be checked against the shipping documents.

Empty trailers should also be sealed immediately after being unloaded. This practice will discourage the use of such empties to remove unauthorized material from dock areas, and it precludes the necessity of physical inspection of all vehicles leaving the controlled areas, providing the seal and the papers check out.

Seal Security

All seals must be held under the tightest possible security at all times. Unissued seals should be logged and secured upon receipt. They should then be issued in numerical order, and the assignment of each should be duly noted in the log. The seal supply should be audited daily; a careful check to account for each one, issued or not, should be made at the same time. Without such regular inventories the entire system can be seriously compromised. As a further part of such audit, all seals which are damaged and cannot be used, as well as seals taken from incoming vehicles, should be logged and secured until they can be destroyed.

OTHER CONSIDERATIONS

Targets of Theft

An analysis of claims data from the transportation industry shows that there are a few very specific commodities that attract the attention of pilferers and thieves.

Nine commodities (clothing, electrical appliances, automotive parts, food products, hardware, jewelry, tobacco products, scientific instruments, and alcoholic beverages) make up about 80 percent of total national losses. Broken down by industry, clothing represents 35.6% of the losses in trucking and 45.5% in air shipments; electrical machinery and appliances make up 11.8% of the losses in trucking and 15.4% of the losses in air transport. These two products alone represent approximately half of all the losses in both air and truck transportation. Obviously a company, carrier, or terminal operator would be well advised to take special precautions when handling such merchandise.

The complete list of principal target merchandise follows:[21]

Truck

1.	Clothing, textiles	35.6%
2.	Electrical machinery, appliances	11.8%
3.	Metal products, hardware	8.9%
4.	Transportation equipment, motor vehicles	5.9%
5.	Food, food products	5.6%
6.	Chemicals, petroleum, rubber, plastics	5.0%
7.	Alcoholic beverages	4.4%
8.	Tobacco products	3.7%
9.	Wood products, furniture	3.6%
10.	Medicine, drugs, cosmetics	3.6%
		88.1%

Air

1.	Clothing, textiles	45.5%
2.	Jewelry, coins	21.0%
3.	Electrical machinery, appliances	15.4%
4.	Machinery, except electrical	3.6%
5.	Instruments	3.5%
		89.0%

Rail/Maritime

1.	Transportation equipment, motor vehicles	28.0%
2.	Food, food products	22.0%
3.	Chemicals, petroleum, rubber, plastics	11.0%
4.	Metal products, hardware	8.1%
5.	Electrical machinery, appliances	8.1%
6.	Wood products, furniture	4.5%
7.	Machinery, except electrical	4.3%
8.	Alcoholic beverages	4.0%
9.	Tobacco products	4.0%
		94.0%

Such a list can be a guide to security managers handling merchandise that falls into one of these categories. This tabulation of industry-wide losses provides a forewarning of the relative dangers.

But the kinds of things stolen vary considerably, depending on the locality and the nature of the merchandise available. A thief might prefer to steal part of a shipment of clothing but, if none is available to him, he will just as vigorously pursue a truckload of dog food, as long as he has a means of disposing of it. Anything can be stolen if the thief is given the opportunity. Anything that has a market—and it wouldn't be shipped unless a market existed—can be considered attractive to a thief.

It is important for transportation industry managers to recognize that all merchandise is susceptible to theft, but at the same time to know the high loss items at their locations so that they may exert extra efforts to secure such goods.

Security Management

Surveys

The security manager of freight terminals or companies engaged in shipping will find himself continually occupied with surveys of the facility under his security supervision.

An initial survey must be made to formulate the security plan governing the premises. It should be thorough enough to detect the smallest weakness in the operation and provide the information needed to prepare adequate defenses. Further surveys will be necessary to evaluate the effectiveness of the program established, and follow-up surveys should determine whether all regulations and procedures are being followed. Additional surveys may be necessary to reevaluate the security picture following changes in operational procedures in the facility, or to make special studies of particular features of the security plan.

Inspections

In addition to these surveys, which are essentially designed to evaluate the security operation as a whole or to re-evaluate it in the light of changing conditions, the security manager should make regular inspections of the facility to check on the performance of security personnel and to check the operating condition of the facility. Such inspections should include potential trouble areas and should not overlook a check of fire equipment and alarm systems.

Education

If the security plan is to succeed, it must have the full cooperation and support of all employees in the facility. This can only be achieved by a continuing program of education in the meaning and the importance of effective security in every phase of the business.

All personnel should be indoctrinated at the time of their employment, and a continuing program should be instituted to update the staff on current and anticipated problems. More advanced courses on procedures might be instituted for management personnel.

Part of this program should be devoted to educating employees in the importance of security to each individual and his job.

Security reminders are also important to keep the subject of security constantly alive in the minds of everyone. Posters, placards and notices prominently posted are all effective devices for getting the message across. Leaflets or pamphlets covering more detail can be distributed to each employee in his pay envelope.

Chapter 11

COMPUTER SECURITY

In the past thirty years American business, in the process of almost explosive evolution, has leaped light years ahead in its capacity to gather, store, and deal with vast amounts of computerized information. This ability has made it possible to effect great improvements in the efficiency and flexibility of all financial matters as well as to create systems providing instant playback describing inventories, market position and potential revenues at any given time.

There seems to be no end to the benefits to business to be derived from computerization, and the more benefits it creates, the more dependent upon its magic we become. In the past decade or so, the computer has become the nerve center of our commercial activity. Indeed, many modern companies are frank to admit that they could not operate without the computer.

A report by the Stanford Research Institute points out that "recent studies indicate that about 7 percent of the present U.S. work force of 84 million people work directly with computers, and another 15 percent work indirectly with them."[22] This same report goes on to say, "All but the smallest business and government agency own, lease or use computer service. Most large organizations are discovering that they can function for only a few hours or few days at most without correct functioning of their computers. At least 60 percent of all banks are automated and would be unable to function unless their demand deposit accounts were successfully processed on computers.

"It is estimated that computer manufacturing, data communications, and operation of computer systems will together represent 14 percent of the gross national product by 1980. A consensus of experts recently indicated that losses, injuries, and damage directly associated with computers will exceed $2 billion annually by 1982."[23]

Computer Vulnerabilities

There can be little question of the importance of the computer in the business world of today and its increasing significance in the future.

Yet in spite of the benefits to be derived from a computerized operation, it creates a great potential for danger. Probably no one element in business, including catastrophic fire and the ravages of non-computerized internal theft, presents a greater potential to wipe out an entire business so quickly and so effectively.

The dangers that can befall a computer center or can be created by it encompass virtually the entire operation of the business it serves. Embezzlement, programming fraud, program penetration, operator error, input error, program error, theft of confidential information, and plain carelessness are a few of the problems that can arise in routine operation. Add to this the potential for fire, riot, flood and sabotage.

The Need for Security

In the face of the risks to this sensitive machinery and to the enormous accumulation of data concentrated in a limited area, as well as business's ever-increasing reliance on the computer, many firms continue to ignore the dangers. Many of them confess an uneasiness about security for the computer but claim they cannot afford an effective protective program.

There is no question that computer security can be costly, but the stakes are too high to try to effect economies in this area. The fact is that any company involved with a computer—and that includes most companies in this country—cannot afford not to have a comprehensive security program that will protect their computer operation. Since such installations can represent an investment running into the millions of dollars, a security program costing from $50,000 to $250,000 is not an unreasonable expense.

Ideally the program designed to provide computer security will be an integrated part of the company's overall loss prevention effort. It should be administered by a specialist in computer operation, but it will report to, and be coordinated with, the larger program. It must encompass physical security, procedural or operating security, access and traffic controls, disaster planning and contingency procedures, employee education programs and a painstaking screening procedure in the employment of computer personnel.

An effective security program is essential to any electronic data processing (EDP) operation, and it must have the vigorous support of an informed management. It's not always easy to convince the chief financial officer of the need to expend substantial funds for security, but when the risks are as high as they are in computer operations, he must somehow be convinced.

PHYSICAL SECURITY

The physical security needs of a computer installation are in most respects the same as any business or industrial establishment. Its goal is essentially to deny access to the computer area to all unauthorized persons.

In many respects this is easier to apply to a computer installation than to other facilities, since only a limited number of people have reason to be in the area in the first place. Whereas an industrial facility may have to make provision for the movement of employees of various trades, truckers, supervisory personnel and others in great numbers, a computer center is not obliged to put up with any appreciable traffic of this kind to operate. It can, and should, be closed to all but those specifically assigned to it.

Structural Barriers

To prevent surreptitious or forced entry, the facility should be housed within an area which is already secured against intrusion or, if possible, in a separate building far enough from other structures to be free from the danger of fire spreading from them. In any event it should be protected by appropriate fencing, whether standing alone or housed within the perimeter protection of another building. It should be adequately lighted and it should not be obstructed by ornamental plantings and landscaping which could provide concealment for an intruder.

Entrances

Entrances should be limited. The best design would incorporate one entrance only and such fire doors fitted with panic bars as are consistent with local fire laws and good safety procedure. Fire doors should be locked, allowing exit but preventing entrance, and they should be alarmed to signal their use.

Any other means of entrance through vents, trash chutes, air conditioning ducts or drains of over 64 square inches should be covered with bars or heavy wire mesh. If the roof can be reached from poles or neighboring buildings, it should be appropriately fenced. If the building is only one or two stories, the roof can be assumed to be accessible and should be reinforced to prevent forced entry by chopping through it. As many windows as possible should be sealed by bricking them up. Those that for any reason cannot be sealed should be covered with a heavy wire mesh of a close enough weave to prevent the introduction of devices such as Molotov cocktails or bombs.

Alarms

Ideally the entire facility should be alarmed, although, if it operates on a 24-hour basis, a full alarm system may not be necessary. If it is not that active, a central station alarm system is essential. The system to be used will depend upon the layout of the facility, but it should certainly cover the entrance, the entrance to the computer itself, and the tape library. In the case of a one or two-shift operation, switch locks which lock off power to the machines should be installed.

Adjoining Spaces

In cases where the computer is located in a building serving other purposes as well, care must be taken to prevent entry into the area by drilling through the floor from areas below or by moving through the crawlspace in the ceiling. Modern construction typically includes space for running cables, air conditioning and other services above the false ceiling. This space is accessible simply by pushing up the suspended ceiling tiles and crawling from office to office. Such a space could provide easy egress to the computer or the computer library unless it is blocked by a bearing wall or by other means.

In this discussion of physical security, we have alternated between assumption of a facility located in a detached building and one located in an existing structure. It might be in either. It would be unrealistic to expect it to be located as sole occupant in a separate structure in an urban setting, but it might be well to consider placing it outside the city. Proximity to the computer is rarely necessary, and the advantages of removing it to a setting where it can be isolated from any but its own problems are considerable. A sizable computer operation represents an enormous capital outlay. It would seem to be poor economy to locate it in a "space available" site in an existing building which is probably poorly equipped to handle the needs of this vital (and expensive) installation.

Site Selection

In considering location, it is important to evaluate the environmental and human factors that might affect the site as well as the services available in the area. Some of the factors that must be considered are:

Fire and Police Protection: Check response capability and general efficiency. In this same regard the availability of central station alarm protection and its reliability must be investigated.

Surrounding Neighborhood: High crime areas are undesirable—not so much because of computer vulnerability as because it is a problem in employee morale. Such a neighborhood could have an adverse effect on employee retention and could also serve to increase insurance premiums in various areas of coverage.

Access: The site must be easily accessible to its personnel. Ideally it should be served by adequate public transportation—remembering that virtually every such installation spends part or all of its time on a 3-shift or 24-hour schedule.

Maintenance: The time of response of manufacturer maintenance and repair personnel must be considered.

Space Requirements: The equipment required for a specific operation may take more space than the headquarters building can provide. The weight of it also may be such that extensive and expensive remodeling may be needed in the existing building.

Natural Phenomena: Consideration must be given to the possibility of floods, earthquakes, hurricanes, tornadoes, etc. Obviously such disasters cannot in themselves be anticipated, but their incidence in the planned site can be. Earthquakes can be anticipated on or near existing faults, floods can be considered a possibility in certain known areas. Some areas suffer from tornadoes or hurricanes more than others. All of these factors must be considered before a site is finally determined upon.

This evaluation should also take into account the presence of radar installations. In the past several years there have been several cases of tapes being erased by radar radiation as much as 1/8 of a mile away.

Power Source: Check the power company's record of reliability and its reputation for efficient and speedy response to power failures. Any fluctuation in line voltage can cause inaccuracies in data transfers. While some operations are not seriously hampered by fluctuations, with others the problem can be a serious one. Where this is the case, the practice is to isolate the computer from the power from the local utility by the use of batteries, constantly recharged by line voltage, which supply the computer through alternators.

Site location must also take into account the presence of water mains and proper drainage in the computer room and library if an already existing building is under consideration. Much damage has resulted from flooding caused by broken water mains or pipes and further complicated by the lack of adequate drainage.

Fire

EDP centers are particularly vulnerable to fire. The maze of wires and equipment under the floor and the accumulation of paper from print-outs, in addition to the normal fuel load characteristic of any office, make for substantial quantities of combustibles to feed any fire that might get a start. The losses in such a fire usually are tremendous. The equipment involved is extremely costly, and the loss of data on destroyed tapes can be many times more harmful and expensive. RCA recently estimated it would take 9.2 man-years and 478 computer hours to reconstruct the master files of its New York City EDP center.

Causes of Fire

The cause of fires in EDP centers is a matter of some dispute. Potentially it would appear that the principal cause should be from electrical malfunctions. The miles of wiring and the possibilities of short circuits are powerful persuasions to exercise great care and to conduct frequent inspections in those areas accessible to inspection.

Statistically, however, the major cause of fire is from adjacent occupancies—either in the same or neighboring buildings. Most EDP centers have been constructed and equipped to resist fire very effectively, but they are often located in a building which is not so well protected. Obviously the computer will go down with the rest of the building in case of fire no matter how well it is itself protected. There have been cases where the floor has collapsed in a fire and dropped the computer and its attendant equipment into the fire below.

Careless use of cigarettes around accumulated waste paper is the second most common cause of fires in EDP centers. Electrical fires are the third most common, but, although it is third on the list, the threat of fire from the power system is always present.

Sensors

Since polyester-backed tape used in computers can be rendered unreadable at temperatures of 150°, it is essential that a sensor system providing the earliest possible warning of a fire be installed. The most economical installation of such a system is during the construction of the facility, but it must be done whether the facility is already built or not. Sensors should be located on ceilings, under raised floors, inside false ceilings, in air ducts and within the equipment itself. This same protection must be extended to the tape library as well. The earliest warning comes from ionization or products-of-combustion detectors, which respond to the earliest stages of combustion before the development of smoke or flame. These sensors must be checked regularly to be certain they are in working condition.

Sprinkler System

Perhaps the most effective automatic fire extinguishing system is general use is the sprinkler system, but its use in computer installations is a matter of considerable controversy. The Factory Mutual group of insurance companies favors the use of sprinklers in computer installations—and they are experts in the field of fire protection. IBM, the world's largest manufacturer of computers, has said that they are harmful to the equipment and should never be used. There are compelling arguments for both views.

Water from a sprinkler system will run indefinitely after it has been triggered by an alarm or by heat, until it is turned off manually. This continuous stream of water cools the burning material below the point of combustion so that it will not start again after the water is turned off. This cooling effect can be important in cases of smoldering combustion (as opposed to flaming combustion).

Unfortunately, there are serious side effects in a computer application. If the power to the computer is not turned off before the water hits the equipment, the result can be ruinous. Electrical circuits short out and the machine may be very seriously damaged. Then, too, tapes can be destroyed. Tapes are not at all affected by a short immersion in water, nor are they damaged by temperatures as high as 250° in a dry environment, but they can be ruined by a combination of heat and humidity. As noted above, they will deteriorate to unusability at 150° in a humid environment. Thus the problem with sprinklers. When they discharge into a fire, they produce steam—which will quickly wipe out the tapes subjected to it.

There are answers to these problems. One lies in the installation of a system which alarms first and provides enough of a delay before water is discharged to allow an employee time to turn off the equipment before water is discharged from the sprinklers. This procedure could also be automatic, whereby the equipment was inactivated by the alarm prior to the flow of water. In such a case the equipment would have to be dried out before it could be used again.

In cases where the facility is shut down, all equipment should be covered to protect it from the possibility of such water damage. If a fire started inside one of the machines, it would burn away the cover and become accessible to the extinguishing efforts of the sprinkler, but the other unaffected machines would still be protected by covers.

Halon Systems

On the other hand, halon 1301, an inert, odorless gas, is an efficient extinguishing agent recommended by many authorities in the field. This gas is non-conductive, leaves no residue and is highly effective in fighting fires. It presents problems, however. It has no cooling effect, and although it is extremely effective when used against flaming combustion in solid fuels, it is considerably less effective against smoldering combustion. Every automatic halon system is limited by the quantity of halon contained in its storage receptacle. Whereas water will continue to flow until turned off, the halon will be discharged only until it is exhausted. Usually, this will be enough, but in some cases of stubborn smoldering fires it may not be.

Finally, whereas halon in its normal use presents no hazard to humans, it is a toxic chemical. Although the National Fire Protection Association has stated it has no effect on humans in concentrations up to 7 percent, it does have harmful effects in concentrations beyond that. They further suggest that concentrations of from 15 to 20 percent might even cause death if exposure to such concentration is prolonged.

Carbon Dioxide

Carbon dioxide systems are as effective as halon, but CO_2 is very dangerous to use since it is deadly in the concentrations needed. All personnel must be evacuated from the immediate premises before it can be discharged. If a CO_2 system is installed, all personnel must be trained in evacuation procedures and indoctrinated in the dangers of this gas.

Dry Chemicals

Dry chemicals are effective in smothering a fire, but they leave behind a residue of powder on the equipment and in the air which may be more harmful to the machines than the fire itself.

Extinguishers

CO_2 extinguishers should be placed throughout the facility and all personnel should be schooled in their use. These extinguishers are not a substitute for a total extinguishing system, but they are valuable for bringing small fires under control before they become big enough to require disruptive and expensive discharge of the main system.

General Fire Prevention

Perhaps the most important element in the fire program is the creation of a carefully supervised fire-resistant environment. This involves a trash disposal system which will prevent the otherwise inevitable build-up of waste paper in the computer area; the establishment of no smoking rules in and around the computer room and the tape library; frequent inspections and clean-ups under the raised floor; fire-proofing all rugs, drapes, wallpaper and furniture; sealing off all holes and passages that violate the fire-resistant integrity of floors, walls and ceilings; installing multiple manual cut-offs of air conditioning.

Access Control

Since one of the principal means to computer security is access control, it must be planned so that those employees authorized to be in the EDP center can get in with a minimum of restrictions and red tape, while unauthorized persons are positively excluded. There are a number of systems currently in use that accomplish this with a high degree of success.

Entrances

There should be an absolute minimum of entrances to the facility. The entrance should be locked at all times. If at all feasible, this entrance should be under the supervision and control of a security employee.

Some installations use a double door arrangement with an electrical control allowing entrance and exit. In this construction, an outside door leads into a small anteroom large enough to accommodate two or three people. Beyond that is another door leading into the facility. When employees arrive they are identified by the guard and allowed through the first door. They remain in the anteroom until the outside door closes and locks before they are allowed in the second door. This arrangement protects against unwelcome persons crowding into the facility along with authorized employees. In larger EDP centers this entire operation can be supervised by a guard observing on CCTV. He activates the electrically controlled doors remotely from his control position.

Locks and Keys

In smaller installations where such an elaborate set-up is not practical, authorized employees may be issued keys to the entrance. This system is generally unsatisfactory, since keys can be lost or compromised.

A better system might be the installation of one of several kinds of combination locks. Combinations can be changed periodically to prevent entry by former employees or other persons who may have discovered the combination. Other locks are available that are opened by the insertion of cards magnetically coded both to identify the user and to open the door. Systems using both the coded card and a combination lock are in effect to avoid the possible compromise of either system above.

There are also locks that identify personnel by their finger-prints. The index finger is placed in a slot and an identifying card is placed in another slot at the same time. A film clip containing the employee's fingerprint is automatically selected from a file, compared with his finger, and entrance is effected. There are even electronic keys that are, in essence, mini-computers, which can be used to gain entrance.

Each system has its strengths and its weakness. No system is infallible, but they should be studied to determine which is the most suitable for a particular installation. Cost is always a consideration, as is the efficiency and dependability of the mechanism.

If any system provides high security at the expense of reason-able convenience to authorized personnel, it may ultimately prove to be totally unsatisfactory. As with any other control, care must be taken in the selection of an access system that it does not create more problems that it solves. If the efficiency and morale of employees is substantially reduced by the installation of any system, it almost certainly should be changed, no matter how security effective it may be.

SYSTEM PROTECTION

Back-Up Systems

Because when a company changes over from manual to compu-ter operation it becomes dependent on the computer, it is vital that some contingency plan be developed for emergencies of any kind that would render the computer inoperable. Such a break-down could come from any of a number of sources—machine malfunction, power failure, natural disaster, fire, malicious de-struction, or even building renovation.

It is customary in setting up such a plan to locate a computer of the same model or conformation and to enter into some kind of mutual assistance pact, in which it is agreed that, if either party should suffer from some kind of breakdown, he can continue to operate in the other's facilities.

Compatibility

This is sensible pre-planning, but it must be pursued much farther than this. It is important, in the first place, to remember that not all machines—even of the same model—are compatible. Systems vary according to ancillary equipment and memory size. Even in cases where the equipment is identical in every detail, differences in programs may create incompatibility.

There is no real way of knowing whether one system will operate with another's software without a trial run. Before any mutual agreement is entered into, it is essential that each test his needs on the other's equipment. If it is determined that compatibility does exist, periodic test runs should still be made to verify the continuing compatibility of the systems. Since equipment and program changes are common in computer operations, there is always the distinct possibility that what is a suitable back-up system today will be unusable when it is needed most. Only regular testing can establish this. Agreements to notify the other of any changes in the system cannot be depended upon as reliable.

Another problem that arises regularly in such agreements is the change in each party's requirements for machine time. If both concerns are on a 16-hour machine use schedule, they have a basic problem to begin with. Even if they are each on an 8- or 12-hour schedule at the time the agreement is entered into, there is no assurance that their needs will not increase to such an extent that neither can fully accommodate the other. Ideally a similar agreement should be made with several companies to assure continued operation in the face of an emergency. This complicates the problem, but it is sufficiently important to make the effort worthwhile.

Off-Premises Protection

However effective security measures may be, there is always the possibility the defenses may be penetrated. If such penetration should result in the loss or destruction of stored tapes, the results could be catastrophic.

Duplicate Data Storage

Here, as elsewhere, it is important to remember that a company, once it has undergone the difficult process of converting its operation to computers, cannot operate manually. Therefore the loss of much of its carefully accumulated data would be ruinous.

The destruction of tapes carrying perpetual inventory information, or a record of accounts receivable, could create problems from which it might never recover. It is therefore essential that duplicates of such data be stored in some remote location where its protection can be assured.

Data stored in this manner must, of course, be regularly updated to reflect the current status of the company. Some of it may be of such a nature that it may need to be revised only once a month, whereas others should be updated much more frequently.

How these duplicate tapes are created is a matter to be determined by the cost factors involved. Though some firms go back as many as five generations in off-site storage, most use a three-generation system, which consists of a duplicate of data up to the time of storage and journal tapes which provide the information to update the latest version. The last full data tape is also stored off-site to be kept until a new updated tape is made. This method of safety storage is more practical than running duplicate tapes of all programs as they are made, although this system is used in some instances. Whether duplicate or journal tapes are used can be determined only by the needs of the company involved. In any case, this protection of irreplaceable data is essential to any prudent security plan.

As a final note it must be pointed out that the destruction of computer data is less often the result of criminal activity than the result of natural disaster or fire or, even more often, program error, machine malfunction or input error. Whatever the cause, however, the results are the same, and some carefully supervised system of duplication must be installed to protect against these results.

Computer and Remote Terminal Access

One of the particularly disturbing problems faced by a company which has converted to a computer operation is the possibility that unauthorized persons might operate the machine either at the installation or through remote terminals, thus gaining access to privileged data which they can then convert to their own use. This is a very real and valid concern, since all the company's files and records have, in fact, been transferred to an instrument that is accessible to many people within the company, or, in the case of time-sharing operations, to many people outside the company.

Passwords

The machine contains a complete and accurate data source essential to the operation of the business it serves. It will, upon command, produce any or all of that information. It will do precisely what it is told to do. Security of data lies in concealing what the machine is told to do. This usually consists of a file system with the establishment of a password when the file is created. The machine is, in effect, told to release such files only after the programmed password is used when the information is requested. It is further instructed never to reveal the password in readouts or system reviews.

Graduated Access

Some files are further protected by a pre-programmed permissions system in which the names of authorized system users are entered into the file on graduated access basis. In this way, a file can be used to its fullest extent by allowing full and restricted use simultaneously to users at various levels of authority. Machines whose use is restricted to only a few special employees are seriously hampered in their effectiveness. A system of graduated access, however, opens up its potential usefulness and at the same time restricts its use to a need-to-know or use basis.

Remote Terminals

In the early stages of computer technology, information was produced only at the site of the machine itself. Protection of its information was simpler, since denial of physical access to the machine denied access to its information. This is not to say that unauthorized personnel did not obtain sensitive information directly from the computer, but when this occurred it was because of a failure in the personnel access procedures.

Today, however, thousands of concerns operate on what are referred to as management information systems. Such systems, much like airline reservation systems, locate consoles, teletypes and other readout devices in convenient locations within the company building or in a network of locations tied in with telephone lines across the country. Many of these devices enable a user to access the computer. In the same way, smaller companies using a computer on a time-sharing basis have similar terminals with which they can access the computer operated by a service bureau.

In both cases the security problem is the same, though its focus is somewhat different. In the case of the proprietary computer, it is necessary to restrict information to those people and those locations where such information is authorized. The time-sharing customer must be restricted to access to his own data base. He must be prevented from entering the files of other customers. How he restricts that information within his own company is for him to decide and program accordingly, but he, as well as all other customers, must be assured that his file is secure from access by any but authorized personnel.

In either system, both personnel and location authorization must be considered as part of the security system. For example, the personnel manager may be authorized to retrieve payroll information at a location within the confines of the personnel department. Although he is authorized to receive the data, it would be denied to him if he asked for it from any other location. This is an important precaution in areas where a visual display might well expose confidential information to any number of unauthorized people.

Checks and Audits

Examining the total process of information retrieval in a security situation, we would find that an initial check and ongoing audit of the on-line user's authority is basic to the system. The user would activate a remote station, at which time he would be asked for his name, organization and password. The software would identify the individual, verify his authority, acknowledge the password and combine that verification with the status of the terminal in use and the access level of the user. After this initial access, the user would ask for a particular file, using such other passwords as may be indicated. Upon further verification the computer would provide him with the data he needed. Any information sought from the machine would be validated before delivery of such information. If, during this time, at any stage, the user attempted to retrieve unauthorized information, appropriate security alarm procedures would be initiated.

SECURITY PROCEDURES

In transferring its operation from manual to computer, a company engages in a difficult, time-consuming job which, without careful pre-testing, may lead to incredible difficulties bordering on chaos. And converting from a computerized operation back to manual is even more difficult. Effectively, once a company has made that decision to computerize, the die is cast. Retreat is virtually impossible.

Thus begins another company's dependence on modern technology. And with this technology come great benefits in time saving, accuracy, data collection and evaluation, invaluable assistance in planning, and instant updated information, to mention only a few of the dividends the computer can so bountifully pay.

But with these benefits come program inaccuracies and fraud, operator errors affecting entire functions of the company, and opportunities for virtually undetectable embezzlement as a few of the possibilities. The list is endless and each one can create havoc.

Errors

It is important to review the operation of a computer system in order to grasp the significance of even the smallest error, whether caused by carelessness or by the intent of someone to compromise the machine.

In the first place, computers store information in their own language and there are many different languages available for machine use, depending on its application. Print outs are frequently in direct statements easily comprehended by the uninitiated, but the stored information is in the form of digits. Everything has been turned into a number and, in order to retrieve it, that number must be called for. If anything has in any way compromised the integrity of that number, the information cannot be retrieved. If, for example, some item has been designated a three but, by error, has been entered into the machine as a two, it may never be found. Since the program clearly carries it as a three, there is no way of knowing that it has somehow found its way into the data bank as a two.

It is essential, therefore, that procedures of operation as well as programming include many methods of cross-checking, and the computer must be further programmed to notify the operator if the cross-checks fail.

Qualified Personnel

An insufficiently trained or unqualified operator may not

recognize such problems when they arise and may continue to damage the stored information in such a way that he compounds the original error and creates a chain of lost information. Such a situation can be extremely damaging to an otherwise efficient operation.

This kind of loss of control over the information stored is one of the costliest and most frustrating problems facing EDP managers today. The damage potential from unqualified personnel who, though sincere, are nonetheless quite capable of bungling a job, is a greater security hazard in most companies than all the criminal conspiracies combined.

It is essential that only highly qualified personnel be permitted to function in areas affecting the accuracy and efficiency of the computer operation. To operate otherwise is to invite disaster.

Prescribed Systems of Operation

It is equally essential that systems be established for every part of this invaluable function. A computer is very costly to install and operate. The demands for its use grow as its usefulness becomes more evident throughout the company. Economy dictates that it be used as efficiently as possible.

Obviously, priorities will be established governing the programs to be run, and explicit systems and procedures regarding the operation of these programs must be developed, communicated to all employees involved, and prominently posted in an appropriate place within the computer complex. These instructions should be reviewed regularly for update.

It is frequently the case that procedural rules are developed, distributed and posted and then not pursued. Inevitably operators work out short-cuts which may or may not be beneficial. It is management's responsibility, however, to see to it that all procedures are followed as published.

It must also be remembered that some procedures can be burdensome and inefficient. If such appears to be the case after reasonable re-evaluation, they must be changed. Operators will frequently find simpler and safer ways to achieve an end. Their suggestions can never be ignored. If they come up with a new way to accomplish a job, it should be examined and, if it is feasible, adopted as a change in operating procedures. The important thing is to insist that prescribed routines be followed to the letter, but at the same time to keep an open mind to suggestions for better ways.

Embezzlement

Most operating routines are designed to avoid mistakes that will result in a loss of control over the retrieval of information, or to avoid operational missteps that could damage or destroy data already accumulated. These are the computer professional's nightmare. But it has become increasingly evident that controls must be established to prevent the use of the computer for criminal acts.

Difficulty of Detection

The computer presents a formidable challenge to the security man. Its very efficiency makes its use as an instrument of embezzlement almost impossible to detect—most of the instances of such use that have been uncovered have been as a result of some unforeseen accident, not from routine or even special audits. The revelations yet to come may be astonishing. In fact, Joseph J. Wasserman, who heads a Bell Telephone Laboratories task force seeking to devise methods of auditing computers used by the Bell System, suggests that many companies already have been hit by heavy losses but don't know it. He predicts that within a few years someone will uncover a computerized embezzlement that will make even the $150 million salad oil swindle seem puny.

Computers are operating faster and faster and producing fewer and fewer of the printouts that auditors and financial officers need to follow the flow of dollars processed by the machines. "If auditing staffs don't get involved in designing computer systems soon, they might as well climb up on their stools, pull down their green eyeshades and pray for early retirement to come," says Mr. Wasserman.[24]

Sophistication in computer disciplines is necessary to use the machine for one's own purposes. Fortunately this knowledge is not possessed by every thief, but the conditions are ideal for the thief educated in the ways of this technology. Computer records present a different world from the old-fashioned ledger. There are no erasures—only removals which leave no trace. No numbers or entries need to be doctored. They are simply changed, and it can be done in seconds. With most systems such a change can never be detected.

Almost none of the major embezzlements uncovered were

found by audits of programs, but were discovered only when a check was accidentally returned by the post office, or an accomplice informed the police for revenge, or when an account was audited off schedule. In other words, the result of the rigged program instruction was stumbled upon—in many cases years after systematic embezzlement had begun—but neither the machine nor its software were set up to frustrate these schemes or to detect them after they were established.

Remember—a computer can do only what it is told, and it does that incredibly well. It makes no judgments and asks no questions. It simply performs as directed.

There is not, at this time, a method which is totally embezzler-proof. That day may come, but the state of the art today is such that, if it can be done at all, the machine is essentially inoperable. For all practical purposes, it is a hazard to which every computer operation is heir.

Crime Protective Measures

There are, however, some procedures which, if developed and supervised, will substantially reduce the danger and make a really determined thief work much harder for his loot.

First—Reduce or eliminate contacts between the four principal categories of computer personnel—the programmers, the operators, the test and auditing personnel, and the maintenance crew. C.F. Hemphill and J.M. Hemphill, in their informative book on computer security, *Security Procedures for Computer Systems,* refer to a major Eastern banking chain that has discouraged contact between operators and programmers (whose collusion would be the most likely and the most effective from a criminal standpoint) by "requiring these two classifications of employees to enter widely separated entrance gates, to park in segregated lots, and to enter work areas through entrances on opposite sides of the building. Continuing this separation, cafeteria access, coffee breaks, and toilet facilities are provided in different areas of the building. Passing from one area to the other with a written pass is possible only during hours of controlled supervision. At other times an alarm system and uniformed guards prevent traffic between programmer areas and operating areas."[25]

These are extreme measures, to be sure, and perhaps more stringent than is necessary or even possible from a labor relations view in other applications. But the point is clear. Programmers must not be permitted to be involved in machine operation—not personally nor through collusion with an operator. If such a situation were permitted, any program man who was criminally inclined could build any loophole he liked into the system and then feed it any information necessary to accomplish his purpose.

For much the same reason, though to a lesser degree, contact between programmers, operators, test and audit personnel, and maintenance personnel should be minimized. It must be remembered, however, that employees in all of these categories are professionals in a complex and exacting field. They cannot be treated arbitrarily, simply because they probably won't put up with it; and they cannot be hampered or tied up in excessive red tape or their usefulness and efficiency will be substantially reduced.

Second—Insist that all programs be documented as the program is being developed. Frequently programmers fail to record their progress or even record the details of the program after it has been completed. This can lead to some confusion and certainly loss of time when errors need correction or changes need to be made in the program. This oversight is also a handicap to new men assigned to the project, who have no way of knowing what the program consists of either up to the stage when they are assigned or, in the case of a completed program, in its final form.

Third—Insist that programmers stay out of the computer room. They rarely if ever have a function to perform there, and their presence could lead to problems for the company.

Fourth—Transfer programmers and operators to different programs and different machines from time to time. This may serve to discourage any plans to set up long term programs of embezzlement. More often than not, the schemes that have been discovered involved the regular issuance of checks or a regular transfer of funds rather than a single raid on company assets. Such transfers could minimize the risk of such steady drains on company funds and increase the likelihood of discovery of any plot by the newly assigned operator or programmer.

Fifth—Control all aspects of machine operation. If an operator has free access and control over input operations, he has effectively become a programmer-operator and can, on his own, use the machine in a number of unauthorized, and possibly criminal, ways. The status of any program run should always be known and its progress supervised.

Sixth—Insist on careful logging of all aspects of the operation. If trouble develops in the operation, reference to such a log, if well and accurately kept, can serve to locate the source of the difficulty.

Seventh—Separate computerized check writing from the source authorizing their issuance. This is only reasonable business practice which would be demanded in any company—computerized or not.

Eighth—Insist on careful audits and evaluation of machine usage. Such an examination will keep management abreast of computer needs and activities for consideration in assigning priorities of use. It will also help to discover unauthorized or unnecessary usage of this valuable instrument.

Ninth—Establish a routine for the regular disposal of all output information, punch cards, program information, etc. This material should be shredded or carried off under supervision to be incinerated.

Tenth—Check on programs periodically after adoption. Such checks can be made by comparing the original copy of the program against the copy being used by the operator. If any changes have been introduced, they should be immediately apparent. No changes should be permitted without supervisor approval.

Eleventh—Test-run all new programs thoroughly before allowing them to become operational. Without such pre-testing undertaken under a prescribed routine, some programs can create havoc with customer relations or they can damage existing data.

Personnel

In the last analysis, no system of safeguards can absolutely prevent theft by computer. There are too many people who must have access to programs and the machine itself in order to function. Any one or group of them could turn their expertise and virtually private knowledge of the company's systems into a criminal scheme. No company can ultimately protect itself against EDP personnel by guards, access systems or even validation systems. Such protective devices can keep out strangers or dabblers, but not the authorized personnel. They create and operate the systems, and they can destroy them.

This is by no means to say that they are, as a group, inclined to damage the company. Quite the contrary, they are usually dedicated specialists whose instinct is to always improve existing operations by creating foolproof and tamper-proof programs. And they should be recognized as such. Security systems which materially reduce their efficiency will be resented, although they, as a group, are inclined to be the first to recognize the need for reasonable security procedures.

Screening and Evaluation

Careful screening of personnel is of the highest priority in any computer security program. Exhaustive evaluation of background, previous employment experience, if any, and level of training and competence is a must. Having satisfied this basic requirement, every new employee must be thoroughly indoctrinated in the peculiarities of the company's operation. He is then phased into an operational function, but without undue haste. This phase-in period will permit management to further evaluate his ability and attitude.

It is also important to remember that the evaluation at this phase is mutual. The new man will be weighing the acceptability of his new position to him with just as much thought and concentration as the company will be determining his suitability to its operation. Since recent studies have indicated that the average annual turnover in computer personnel amounts to approximately 15 percent, and competent trained people are in short supply in this exploding market, it is important to keep this mutuality of satisfaction in mind.

A Special Breed

At this point it would be well to consider that computer people are a special breed of men and women. They speak a different language and, generally speaking, their dedication is to the demands of the machine, not to the company that owns it. They pursue challenges wherever they may arise. Encouraging company loyalty is an important consideration and should be pursued in a mature and intelligent manner. The computer room is the last place for locker room pep talks.

Management must recognize that the demands of this technology introduce different work routines, and any attempt to force non-applicable company policies in this regard on computer personnel will be non-productive and probably damaging. This is not to say that they need to be coddled or accorded any special privileges; it is simply to point out that they are specialists in a specialized field, and many of the standard rules and routines are irrelevant and don't apply.

Salaries

They must be adequately paid. This will probably adjust itself, since the demands for good people in this field are so great that offers below the going market will be ignored. But pay scales must be regularly reviewed. If the salary range in a company freezes at a point which is later exceeded by the computer business as a whole, large scale resignations can be expected. Excessive turnover of such personnel, beyond the already considerable rate, is expensive and counter-productive.

If all this seems excessively conciliatory, that's unfortunate. Those are generally the facts. Computer people today are in some respects the glamor element—the stars of the business world. They are now, and for the immediate future, in a position to demand respect and appreciation. Any other attitude by management would be to deny the realities in today's commercial world.

The best security lies in enlightened employee relations combined with conscientious leadership—and the computer staff can be the most important weapon in the fight against computerized crime.

Insurance

Since the potential for loss in a computer operation is so enormous, and since no existing system can provide a guarantee against such losses, insurance coverage is essential to back up the operation after all other reasonable precautions have been taken.

Because this kind of coverage is relatively new in the field of casualty insurance, new policies are still being developed. And since data on liabilities and claims is necessarily limited at this stage, rates may vary considerably from company to company, depending on exposure and operational techniques.

At the moment more claims are presented for damage from

water damage than any other type of loss, but this may not continue as coverage becomes more comprehensive and auditing techniques become more sophisticated.

Equipment Insurance

Equipment insurance is perhaps the first consideration. It covers all equipment for a wide variety of risks from fire to accidental damage. Since such insurance may be offered with a deductible feature, savings may be effected in certain instances. In the case of leased equipment, the contract must be studied to determine the liabilities of lessee and lessor, since some agreements do not hold the lessee liable for damage, while others do in certain instances.

Data Insurance

Software can be covered under the company standard fire insurance contract, but such insurance covers only the physical material such as tape. A separate policy or an endorsement on the standard policy will be required to cover the cost of reconstructing records destroyed by various hazards. Such a policy probably will not cover operator errors which result in damage to the data base.

Other Coverages

Business interruption and extra cost insurance covering computer operations is available and recommended.

Accounts receivable insurance is available, but may not be necessary if a good program of off-site storage is followed.

Errors and omissions insurance could be very important to the operator of a data processing service. This covers liability incurred by such a service making honest mistakes in the performance of a job for a client firm. Recent court cases, in which defense disclaimed responsibility for damages because "it was the computer that did it," seem to indicate that the courts do not consider computer error to be a mitigating circumstance or a valid defense. Insurance coverage in this area is advised.

Many of the above may be covered by an all-risk policy which provides comprehensive coverage of most phases of the computer operation, but only a very careful study will determine its practical application to any given business.

Chapter 12

SECURITY
SERVICES

There is little question that the private security industry offers a wide variety of career opportunities. An increased public awareness of the necessity of protective and equipment services to augment the activities of public law enforcement has led to a steady growth in the private sector—from 1958 to 1968 at a regular compound annual rate of approximately 11%, or from about $1.2 billion to $2.9 billion in annual expenditures. Current expenditures amount to about $5 billion for private security, and the needs in 1978 are estimated at approximately $6.4 billion. Predictions are that, although the *rate* of expansion will decline somewhat, the *growth* will continue.

An interesting aspect of this expanding market is the indication that the percentage of employment of in-house or proprietary guard personnel is declining slightly, while contract security services show a sharp upward trend in employment.[26]

As we have seen, during the 19th and early 20th century public police operated only on a local basis. They had neither the resources nor the authority to extend their investigations or pursuit of criminals beyond the sharply circumscribed boundaries in which they performed their duties. When the need arose to reach beyond these boundaries, or to cut through several of these jurisdictions, law enforcement was undertaken by such private security forces as the Pinkerton Agency, railway police, or the Burns Detective Agency.

As the police sciences developed, public agencies began to assume a more significant role in the investigation of crime and, through increased cooperation, the pursuit of suspected criminals. Concurrent with this evolution of public police, private agencies shifted their emphasis away from investigation and toward crime prevention. This led to an increasing use of guard services to protect property and to maintain order. Today, in terms of numbers, guard forces are by far the predominant security service supplied.

CONTRACT SECURITY SERVICES

Contract security as an industry consists of approximately 4,000 establishments providing guard, investigative, central station and armored car service. Of these, six large publicly owned firms dominate the industry, accounting for approximately half of all revenues generated by contract security services. However, since the total expenditure for contract security by American business amounts to about $1.6 billion, there is still a significant market left for the smaller firms, and opportunities for careers in this area of security are growing every year.

Many firms, particularly the smaller ones, specialize in the type of services offered to a client; the larger the firm, the more apt it is to provide a full range of security services as necessary.

The major categories of these services are guard forces, patrols, private investigation, alarm response, and armored car delivery service.

Guard Service

Guard supply represents the major service provided in the industry today. Only part of a guard's job is crime-related. Whenever possible or necessary, he is required to prevent crimes and to report those that have been committed, but his major role may be to direct traffic, to screen persons desiring access to a facility, and generally to enforce company rules. In many modern applications, his role is less regulatory than helpful. He may direct or escort persons to their destination within a facility; he may act as a receptionist or as a source of information; or he may be primarily concerned with safety.

Since many guards are concerned such a small percentage of the time with crime-related activity, there is some effort in various quarters to adjust the guard's appearance to fit his role, by outfitting him in blazer and slacks rather than the uniform with its police or coercive connotation.

A guard is, however, a guard. Even if, in his particular assignment, he is never confronted with criminal activity, he is still charged with certain responsibilities in that area, and he is additionally responsible to protect the interests of his employer on the employer's property.

Difference from Police Function

Private guards differ from public police both in their legal status and in that they perform in areas where the public police cannot legally or practically operate.

The job of the private guard is to provide specific services under the direction and control of a private employer, one who feels that he needs to exercise controls or supervision over his own property or goods, or to provide additional services that the public police as a practical matter simply cannot provide.

The public police have no authority to enforce private regulations, nor have they the obligation to investigate the unsubstantiated possibility of crime (such as employee theft) on private property. As the crime rate climbs, so are the resources of the public police stretched to a point where they can be used less and less for the prevention of crimes against property. As a result, businesses desiring more or different kinds of service or protection must turn to private security service.

Patrol Service

Private patrol services offer a periodic inspection of various premises by one or more patrolmen operating either on foot or in a patrol car.

The tour of such patrols may cover several locations of a single client, or it may include several establishments owned by different clients within a limited neighborhood. Inspections of patrolled premises may be visual perusal from the outside, or they may require entering the premises for a more thorough inspection. Normally the arrangement made with a client specifies that a certain number of inspections will be made within a given period of time or with a specified frequency.

The patrolman differs from the guard in that he operates through a tour covering various locations, whereas the guard stands a fixed post. The patrol service is more economical, since the guard mans a post for the full period during which a danger exists. But the patrol has the possible disadvantage of being circumvented by an intruder who knows that there will be some period of time between inspections of a given premise.

Private Investigation Service

The private investigator is a gatherer of facts—in essence, a research man who spends the greatest proportion of his time in collecting background information for pre-employment checks on personnel, background checks of applicants for insurance or credit, and investigation of insurance claims. Much of his work is non-crime-related, although he may become involved either part or full-time in undercover investigations of employee theft or in detecting shoplifting.

Investigations in divorce-related matters are declining as divorce laws are becoming more liberalized. Tracing missing persons or investigating criminal matters on behalf of the accused are a very small part of a typical investigator's work. In fact, the Pinkerton Agency will not handle any matter involving the defense of persons under prosecution by the public law enforcement agencies, nor matters pertaining to domestic or marital investigation.

Although there are situations, such as the long-term relationship between the American Banking Association and the Burns Agency, where private investigations are called upon to *supplement* the work of public police, the great majority of private investigative work is *complementary* to the public law enforcement effort.

Alarm Response Service

Central Station alarm systems consist essentially of alarm sensors located in the protected premises, and a communication line from the sensors to a privately owned central station alarm board which is monitored and responded to by private security personnel.

In some cases central station systems do not dispatch personnel to respond to an alarm, merely relaying the alarm received to public police headquarters. But in most such systems the alarm is relayed and someone is also sent to the scene.

Alarms connected to a central station are usually designed to detect intrusion, but they can also be used to monitor industrial processes or conditions.

Certainly central station coverage of a facility is cheaper than full-time security employees performing essentially the same function. A drawback is that the current false-alarm rate for virtually all intrusion systems is 95% or more, resulting in a growing resistance to the use of direct connection systems to police head-quarters. Central station operators have some flexibility in check-ing the validity of an alarm before notifying the police, so they may to some degree reduce the incidence of false alarms demand-ing police response.

In cases where central station personnel actively investigate the intrusion, even taking steps to apprehend a suspect before the arrival of the police, they *supplement* the public police effort.

Armored Delivery Service

Armored delivery services provide for the safe transfer of money, valuables, or any goods the employer may wish to move from one location to another. By far the widest use of this service is to transport cash and negotiables from a receiving point to a bank or other depository. Payrolls, cash receipts, or cash supplies for the daily business are the principal traffic of the armored delivery service.

Personnel employed by such services are not concerned with the general security of the premises they serve; their responsibility is confined to the safe transport of sensitive items as directed by the customer.

Contract vs. Proprietary Services

Since the functions described, with the exception of armored delivery service, can be undertaken as proprietary or in-house activities, why then is there a clear trend toward the use of con-tract protective systems?

The Rand Report previously mentioned discusses the pros and cons of the issue in a presentation which is a synthesis of views from both in-house and contract services, various major users of such services, and material from several periodicals dealing with the subject. Some of their conclusions are reflected in the follow-ing discussion of the relative merits of the two approaches to security services.

Various factors will determine the decision to hire in-house or contract guards in any given situation. These factors will include the location to be guarded, the size of the force required, its mission, the length of time the guards will be needed, and the quality of personnel required.

Advantages of Contract Services

Cost

Few experts disagree that contract guards are less expensive than a proprietary unit. Contract guards typically earn from $1.60 to $2.50 an hour, whereas in-house guards typically earn more because of the general wage rate of the facility employing them. In many cases that wage level has been established by collective bargaining.

Contract guards receive fewer fringe benefits, and their services can be provided more economically by large contract firms by virtue of savings in costs of hiring, training, and insurance because of volume. Short-term guard service on a proprietary basis can create such large start-up costs that the effort is impractical.

Liability insurance, payroll taxes, uniforms and equipment, and the time involved in training, sick leave and vacations are all extra cost factors that must be considered in establishing a proprietary force.

Administration

Establishing in-house guard service requires the development and administration of a recruitment program, personnel screening procedures, and training programs. It will also involve the direct supervision of all guard personnel. Hiring contract guards solves the administrative problems of scheduling and substituting manpower when someone is sick or terminates his employment.

There is little question that the administrative chores are substantially decreased when a contract service is employed. At the same time, the contracting customer is obliged to check the supplier's performance of contracted services on an ongoing basis, and he must additionally insist on a satisfactory level of quality at all times. To this extent, management of the client firm is not totally relieved of administrative responsibilities.

Manpower

During any periods when the need for guards changes in any way, it is necessary to lay off existing guards or take on additional manpower. Such changes may come about fairly suddenly or unexpectedly.

In-house forces rarely have this flexibility in manpower. If they have extra men available for emergency use, such men are an unnecessary expense when they are idle. Similarly, if there is a temporary decrease in the need for guards, it would hardly be efficient to dismiss extra men only to rehire additional guards a short time later when the situation changed again.

With their larger pools of manpower, contract firms can use their personnel with a high degree of efficiency. By proper scheduling they can provide extra men on very little notice. Firms with relatively small needs in guard personnel will find this availability problem significant unless they have access to a ready pool of trained guards. Almost invariably, such a pool is a contract service.

Unions

Guard users in favor of non-union guards support their position by arguing that such guards are not likely to go out on strike; they are less apt to sympathize or support striking employees; and they can be paid less because they receive few, if any, fringe benefits.

Since 90% of all unionized guards are proprietary personnel, anyone subscribing to the arguments listed here would clearly favor hiring contract guards. Only 10% to 25% of the guards employed by the three largest contract guard agencies (Pinkerton, Burns, Wackenhut) are unionized.

Impartiality

It is often suggested that contract guards can more readily and more effectively enforce regulations than in-house personnel. The rationale is that contract guards are paid by a different employer and, because of their relatively low seniority, have few opportunities to form close associations with other employees of the client. This produces a more consistently impartial performance of duty.

Expertise

When a client hires a guard service, he also hires the management of that service to guide him in his overall security program. This can prove valuable even to a firm that is already sophisticated in security administration. A different view from a competitive supplier trying to create good will with his client can always be helpful and illuminating.

Advantages of Proprietary Guards

Quality of Personnel

Proponents of proprietary guard systems argue that the higher pay and fringe benefits offered by employers, as well as the higher status of in-house guards, attracts higher quality personnel. Such employees have been more carefully screened, and they show a lower rate of turnover.

Control

Many managers feel that they have a much greater degree of control over personnel when they are directly on the firm's payroll. The presence of contract supervisors between guards and client management can interfere with the rapid, accurate flow of information either up or down.

An in-house force can be trained to suit the needs of the facility, and the progress and effectiveness of training can be better observed in this context. The individual performance of each member of the force can also be evaluated more readily.

Loyalty

In-house guards are reported to develop a keener sense of loyalty to the firm they are protecting than do contract guards. The latter, who may be shifted from one client to another, and who have a high turnover rate, simply don't have the opportunity to generate any sense of loyalty to the specific—often temporary—client-employer.

Prestige

Many managers simply prefer to have their own men on the job. They feel that the firm gains prestige by building its own security force to its own specifications, rather than renting one on the outside.

CONCLUSIONS

Obviously, in weighing the various factors on either side of this debate, the prudent manager will carefully study the quality and performance of the guard firms available to service his facility. He will assure himself of their standards of personnel, training and supervision. He will make a careful analysis of the comparative costs for proprietary or contract services, and he will make an estimate as to their relative effectiveness in his particular application.

In situations where the demand for guards fluctuates considerably, a contract service is probably indicated. If a fairly large, stable guard force is required, an in-house organization might be favored.

As an indication of the experience of a sample of industrial firms, a survey undertaken by the American Society for Industrial Security revealed that approximately 40% of the respondents used contract services for 50% or more of their guard needs. Of all users of some contract security, over 50% responded that they did so for reasons of economy or to avoid the administrative headaches of labor and personnel problems. Generally speaking, their experience with such service was rated as from fair to good. [27]

PRIVATE SECURITY PROBLEM AREAS

Although the growth of the private security industry clearly indicates that a need for such services exists on a steadily increasing basis, the industry confronts problems that are generally recognized by security professionals as well as by their customers. For various reasons which we shall briefly examine, little progress has been made in finding solutions to many of these concerns. Nevertheless, they will ultimately have to be dealt with if the industry is to move toward professionalism and take the place it should reasonably occupy in the business community.

Personnel Problems

"The typical private guard is an aging white male who is poorly educated and poorly paid. Depending on where in the country he works and on his type of employer (contract guard firm, in-house firm, government), he has the following characteristics: his average age is between 40 and 55; he has had little education past the ninth grade; he has had few years of experience in private security; he earns a marginal wage of between $1.60 and $2.25 per hour and often works 48 to 56 hours per week to make ends meet. If employed part-time, he works only 16 to 24 hours per week. Often he receives few fringe benefits; at best, fringe benefits may amount to 10 percent of wages. Guards have diverse backgrounds; many are unskilled; some have retired from a low-level Civil Service or military career; younger part timers are often students, teachers, and military personnel on active duty. Annual turnover rates· range from less than 10 percent in some in-house employment to 200 percent and more in some contract firms. Few guards are unionized. . . . .

"*In general, public police draw upon a different labor pool than do private security forces, with the possible exception of private investigators and security executives.* The principal differences that lead to separate labor pools are the nature of the work, the levels of wages and fringe benefits, the age and education requirements of public police, and the lengthy screening policies for public police personnel. Only a small percentage of private security personnel have ever applied for a public police job, and former law-enforcement officers seldom switch to non-management private security employment."[28]

Training

All the literature on guard force administration stresses the need for adequate training programs. The authors of the Rand Report, frequently referred to in this section, confirm that, in their contacts with a wide variety of people holding various positions in private security, no one raised the slightest doubt about the necessity for training private guards, nor did they question the existence of significant variations in the quality of existing guard training programs. But there is no discernible concurrence on the amount of training necessary to adequately train a guard, nor is there yet any real groundswell demanding that training—even for armed guards—be mandatory to the most elementary degree.

In a survey of private security personnel undertaken in the preparation of the Rand Report, 65% of the respondents reported receiving no training prior to actually beginning work. Those who did receive some training typically read a manual or were interviewed by a superior. Less than 7% received more than eight hours of initial pre-work training; 19% reported being put to work by themselves on the first day, and the remainder received a small amount of on-the-job training by a superior or fellow employee. It is also interesting to note that this same survey revealed that, while 47% carry guns, only 19% received initial firearms training, and only 10% receive periodic firearms re-training.

It is important to remember that the data above was developed from a limited survey of guard personnel, but it covered a wide representation of the industry, from in-house guard forces in various industries to different types and sizes of contract services. Its conclusions cannot be ignored. Nor can we dismiss the comment of Charles E. Lamb, executive of the United Plant Guard Workers of America, when he stated that "from long experience in representing the agency guard, I can tell you that I have yet to see the guard agency that actually trains guards to any extent at all."[29]

Both the observation by Mr. Lamb and the conclusions of the survey are at some variance with the descriptions of training programs offered by various security executives interviewed for the Rand Report. But even in those descriptions, total initial pre-work plus initial on-the-job training is less than two days in duration for a majority of the private guards working in the United States today.

In many cases, this generally low level of training might lead only to a lower efficiency level of performance by guard personnel. But it could also have serious repercussions involving violation of the rights of citizens as well as civil or criminal liability for wrongful actions by an improperly instructed guard.

The extent of the problem is further emphasized by the Rand Report: "Our survey of security employees. . .also contained several questions to test the guard's knowledge and his reaction in several hypothetical situations. Each employee surveyed was given a total of 44 chances in the questionnaire to make a 'mistake.' Twenty of these 44 potential errors were 'major,' i.e., errors that could result in an improper guard action that might lead to civil or criminal charges. The results were shocking. Over 99 percent of the security workers made at least one mistake; the average was over 10 mistakes. More significantly, over 97 percent made at least one major error; the average was 3.6 major errors, any one of which could lead to civil or criminal charges against the employee and/or his employer! One very reasonable hypothesis is that these men were not well trained. These results are even more significant, in view of the fact that our employee survey was biased in favor of higher-paid, better-educated security workers, who were allowed time to think before responding to the questions. That is, they were not forced to make the decisions in a crisis situation."[30]

Most security executives agree that levels of training in the security business are generally inadequate, though few will acknowledge that their own programs fail to meet the reasonable standards that they perceive to be indicated.

The common explanation for the poor quality of training is the economics of the marketplace. There is general agreement that any contract service that invests in the otherwise non-productive or non-revenue-productive time and effort of training is at a competitive disadvantage when bidding on a job.

It is true that too many firms contract for security services with the low bidder, without reference to the other factors that should properly be considered in assigning such an important contract; but the costs of a training program need not be so great that they would automatically leave the field open to the most price-conscious competition.

The authors of the Rand Report have estimated that the increase in contract fee required to cover a reasonable three-week training program is at the most $.23 per hour per guard, assuming wages of $2.00 an hour, instructor wages of $4.00 an hour, and an overhead cost of 50%, with ten trainees per class, if the average employee stays on the job for one year. A shorter program, larger classes, or a longer average term of employment would all have the effect of considerably reducing the cost.

Abuse of Authority

Inevitably, the lack of adequate indoctrination or training leads to misconceptions of the role and the authority possessed by private security personnel. Apparently many of them are of the opinion, consciously or unconsciously, that they have the same general legal powers as public police—powers which, quite naturally, they understand imperfectly. Consequently, security personnel frequently overstep their authority. It would appear that they are subjected to legal action, civil or criminal, less often than they might be in such cases only because the general public is similarly ignorant of the limitations of the powers of these uniformed officials.

Indications are that incidents of assault, use of unnecessary force, and improper arrest or detention are common enough to be disturbing. And the potential for such incidents is in no way mitigated by an often indifferent reaction on the part of the private security industry to these events.

Professionalism in Security

There is little question that the problems cited here will continue to plague the industry until such time as mandatory training, based upon a widely accepted syllabus, is accepted as a requirement in the security business. Although a few communities and one state (Ohio) have certain training requirements for security personnel, there is not yet a general demand for the establishment of such standards nationwide.

Security today is a growing and generally lucrative field, and there is little inclination in many quarters to disturb the *status quo.* Other responsible security executives, however, have begun to express their concern over the lack of regulation of the industry, a situation that leads to cutting corners and lowering standards to meet the competitive threat.

Need for Regulation

The day may soon be at hand when the need for regulation will be so clearly perceived that it can be effected, to the benefit of supplier, user, and the general public alike. Such regulation need not be totally imposed by government agencies. Indeed, to be genuinely effective it should be self-imposed by the industry. Tighter government controls are clearly indicated, but the thrust toward the achievement of professionalism should properly come from within.

Minimum Standards

There should be a universally accepted code of ethics to establish a high standard of professional conduct for security officers. Every effort should be made to develop procedures, techniques and skills into a consistent body of theory governing the application of security systems. Such theory should be regularly and freely disseminated to all members of the industry.

Most significantly, certain requirements for entry into the field should be established. These should include minimum training, certification of attainment by the industry, and licensing by some regulatory body.

Ultimately, the establishment of an organization that would draw up guidelines, set policy, and sponsor research or act as a central clearing house for the distribution of the results of such research would create the framework for true professionalization of private security, an industry that in so many ways touches the lives of all of us.

The challenge is an imposing one. Because it is, the opportunities are enormous.

FOOTNOTES

1. *An Analysis of Criminal Redistribution Systems and Their Economic Impact on Small Business* (October 26, 1972), pp. 21-29.

2. *Ibid.,* p. 25.

3. U. S. Department of Commerce, Preliminary Staff Report on *The Economic Impact of Crimes Against Business* (February, 1972), p. 22. The figure quoted here refers to estimated expenditures in crime prevention as of 1969. The operative figure of approximately $5 billion used both in this chapter and in Chapter 12 represents the estimated expenditures in 1974 on crime-related security services and equipment, based on an extrapolation of the growth factor in the industry.

4. Specifications of government requirements can be obtained from *Industrial Security Manual for Safeguarding Government Classified Information,* available from the U. S. Government Printing Office.

5. C. F. Hemphill, Jr., *Security for Business and Industry* (Dow Jones-Irwin, Homewood, Ill., 1971), p. 245.

6. Pratt, L.A., *Embezzlement Controls for Business Enterprises* (Fidelity and Deposit Co., Baltimore, Md. Revised Ed. 1966), pp. 30-31.

7. Curtis, Bob, *Security Control: External Theft* (Chain Store Publishing Corp., New York, 1971), p. 12.

8. *Ibid.,* pp. 15-16.

9. *Successful Retail Security* (Security World Publishing Co., Inc., Los Angeles, 1974), pp. 188-193.

10. *Ibid.,* p. 192.

11. *Ibid.,* p. 191.

12.   Curtis, *op. cit.*, p. 65.

13.   *Ibid.*, pp. 75-76.

14.   *Ibid.*, p. 120.

15.   *Ibid.*, p. 110, quoting Karl Menninger, M.D., "Verdict Guilty—Now What?" *Harper's Magazine,* Vol. 219 (July, 1958), pp. 131-143.

16.   *Successful Retail Security*, pp. 44-49.

17.   *Ibid.*

18.   U. S. Department of Commerce, *op. cit.*

19.   Barlay, Stephen, *Fire* (Stephen Greene Press, Brattleboro, Vt., 1973), p. 192.

20.   *Ibid.*

21.   "How, Where, When and What," *Transportation and Distribution Management*(July, 1972),Benjamin O. Davis, Jr., p. 25. Cited in Ursic and Pagano, *Security Management Systems* (Charles C. Thomas, Springfield, Ill., 1974), p. 161.

22.   Parker, Nycum and Oura, *Computer Abuse* (Stanford Research Institute, Menlo Park, Calif., 1973), p. 17.

23.   *Ibid.*

24.   Adelson, A., "Embezzlement by Computer," in *Computer Security: Equipment, Personnel and Data* (Security World Publishing Co., Inc., Los Angeles, 1974), p. 43.

25.   Hemphill, *op. cit.*, p. 113.

26.   Kakalik, J.S. and Wilhorn, S., The Rand Corp., Santa Monica, Calif. Under Grant NI-70-057, Vol. II, R8701-DOJ. *The Private Police Industry,* p. 19.

27.   *Ibid.*, p. 53.

28.   *Ibid.*, p. 67.

29.   *Ibid.*, p. 85.

30.   *Ibid.*, p. 88.

A SELECTED BIBLIOGRAPHY

Bray, Samuel E. and Hurley, Robert. *Freight Security Manual.* Los Angeles: Security World Pub. Co., 1970.

Cole, Richard B. *The Application of Security Systems and Hardware.* Springfield, Illinois: Charles C. Thomas, 1970.

Currier-Briggs, Noel. *Security: Attitudes and Techniques for Management.* London: Hutchinson & Co., 1968.

Curtis, Bob. *Security Control: External Theft.* New York: Chain Store Pub., 1971.

Curtis, Bob. *Security Control: Internal Theft.* New York: Chain Store Pub., 1973.

Curtis, Bob. *Modern Retail Security.* Springfield, Illinois: Charles C. Thomas, 1960.

Gammage, Allen Z. and Hemphill, Charles F., Jr. *Basic Criminal Law.* New York: McGraw-Hill, 1974.

Hemphill, Charles F., Jr. *Security for Business and Industry.* Homewood, Illinois: Dow Jones-Irwin, Inc., 1971.

Hemphill, Charles F., Jr. and Hemphill, John M. *Security Procedures for Computer Systems.* Homewood, Illinois: Dow Jones-Irwin, Inc., 1973.

Hughes, Mary Margaret, ed. *Successful Retail Security.* Los Angeles: Security World Pub. Co., 1974.

Inbau, Fred E. and Aspen, Marvin E. *Criminal Law for the Layman.* Philadelphia: Chilton Book Co., 1970.

Industrial Safety and Fire Prevention. London: T. Bell & Son, 1973.

Kingsbury, Arthur A. *Introduction to Security and Crime Prevention Surveys.* Springfield, Illinois: Charles C. Thomas, 1973.

Mandelbaum, Albert J. *Fundamentals of Protective Systems.* Springfield, Illinois: Charles C. Thomas, 1973.

Momboisse, Raymond M. *Industrial Security for Strikes, Riots and Disasters.* Springfield, Illinois: Charles C. Thomas, 1968.

Oliver, Eric and Wilson, John. *Practical Security in Commerce and Industry.* London: Gower Press, 1968.

Post, Richard S. *Combating Crime Against Small Business.* Springfield, Illinois: Charles C. Thomas, 1972.

Post, Richard S. *Determining Security Needs.* Madison, Wisconsin: Oak Security Pub., 1973.

Post, Richard S. and Kingsbury, Arthur A. *Security Administration: An Introduction.* Springfield, Illinois: Charles C. Thomas, 1973.

San Luis, Ed. *Office and Office Building Security.* Los Angeles: Security World Pub. Co., 1973.

Strobl, Walter M. *Security.* New York: Industrial Press, 1973.

Ursic, Henry S. and Pagano, Leroy E. *Security Management Systems.* Springfield, Illinois: Charles C. Thomas, 1974.

Wathen, Thomas W. *Security Subjects.* Springfield, Illinois: Charles C. Thomas, 1972.

Woodruff, Ronald S. *Industrial Security Techniques.* Columbus, Ohio: Merrill Pub., 1974.

Wright, K. G. *Cost-Effective Security.* (UK) Ltd: McGraw-Hill, 1972.

Other Publications

Applying the OSHA Standards. A manual from National Loss Control Service, Long Grove, Illinois, 1973.

Cargo Loss Prevention Recommendations. International Union of Marine Insurance, Zurich, 1970.

Colling, Russell L., ed. *Hospital Security and Safety Journal Articles.* Medical Examination Pub. Co., Flushing, N.Y., 1970.

Guidelines for the Physical Security of Cargo. Department of Transportation, May, 1972.

OSHA Reference Manual. Insuror's Press, Santa Monica, Calif., 1972.

Pratt, Lester A. *Embezzlement Controls for Business Enterprises.* Fidelity and Deposit Co., Baltimore, 1952.

Security World Magazine. Monthly except July/August issue. Security World Pub. Co., Los Angeles.

Survey sponsored by the National Institute of Law Enforcement and Criminal Justice.

Kakalik, James S. and Wildhorn, Sorrel, authors of The Rand Corporation study under Grant NI-70-057.

R-869-DOJ *Private Police in the United States: Findings and Recommendations.*

R-870-DOJ *The Private Police Industry: Its Nature and Extent.*

R-871-DOJ *Current Regulation of Private Police: Regulatory Agency Experience and Views.*

R-872-DOJ *The Law and Private Police.*

R-873-DOJ *Special-Purpose Public Police.*

Thorsen, June-Elizabeth, ed. *Computer Security: Equipment, Personnel, & Data.* Security World Pub. Co., Los Angeles, 1974.

Uniform Crime Reports. Federal Bureau of Investigation, Washington, D.C.

United States Department of Commerce: Preliminary Staff Report on *The Economic Impact of Crimes Against Business.* February, 1972.

Walsh, Timothy J. and Healy, Richard J. *Protection of Assets Manual.* The Merrit Co., Santa Monica, Calif., 1974.

INDEX

ALARM SYSTEMS & THEFT PREVENTION
By Thad L. Weber (385 pp.)

First definitive treatment, in laymen's terms, of the so-called "impregnable" alarm systems, their strengths and weaknesses, how they are attacked by criminals, and the techniques required to defeat these attacks. A highly readable guide to the entire gamut of alarm problems.

INTERNAL THEFT: INVESTIGATION & CONTROL
An Anthology (276 pp.)

Two dozen top security professionals analyze employee dishonesty and how to control it. 25 chapters on: Why Employees Steal; Executive Dishonesty; Embezzlement; Undercover Investigation; Interrogations; Confessions; Polygraphing; Pre-Employment Screening; Management's Role, and much more!

HOSPITAL SECURITY
By Russell L. Colling (384 pp.)

Complete protection of people and property in health care facilities. Developing a practical, detailed program for establishing a security system or refining existing programs to deal with hospital vulnerabilities, including: theft of drugs, assault, kidnapping, fire, disaster, strikes, robbery, accidents, vandalism, internal theft, more.

HOTEL & MOTEL SECURITY MANAGEMENT
By Walter J. Buzby II and David Paine (256 pp.)

Security hazards in the hotel industry, and protective measures which can help the hotel or motel, large or small, prevent losses. Includes theft, holdup, fraud, fire, disaster, restaurant and bar security, access control. Special emphasis on laws affecting hotels, including innkeeper's liability for injury to guests due to accident or crime.

In addition to its hard cover books on security subjects, Security World Publishing Company publishes *Security World* and *Security Distributing & Marketing (SDM)* magazines; produces booklets, manuals and audio tape cassettes on security; and sponsors the International Security Conference, held annually in Chicago, New York and Los Angeles. Books and other materials are available from Security World Publishing Co., Inc., 2639 So. La Cienega Blvd., Los Angeles, Calif. 90034.